Grade 4

Reveal MATH®

Differentiation Resource Book

mheducation.com/prek-12

Send all inquiries to:
McGraw-Hill Education
8787 Orion Place
Columbus, OH 43240

ISBN: 978-1-26-421064-0
MHID: 1-26-421064-7

Printed in the United States of America.

1 2 3 4 5 6 7 8 9 LHN 25 24 23 22 21 20

Grade 4

Table of Contents

Unit 2
Generalize Place-Value Structure

Unit 3
Addition and Subtraction Strategies and Algorithms

Unit 4
Multiplication as Comparison

Unit 5
Numbers and Number Patterns

Unit 6

Multiplication Strategies with Multi-Digit Numbers

Lessons

Unit 7

Division Strategies with Multi-Digit Dividends and 1-Digit Divisors

Lessons

Unit 8

Fraction Equivalence

Lessons

Unit 9

Addition and Subtraction Meanings and Strategies with Fractions

Lessons

Unit 10

Addition and Subtraction Strategies with Mixed Numbers

Lessons

Unit 11

Multiply Fractions by Whole Numbers

Lessons

Unit **12**

Decimal Fractions

Lessons

Unit **13**

Units of Measurement and Data

Lessons

Unit **14**

Geometric Figures

Lessons

Understand the Structure of Multi-Digit Numbers

Name ______________________________

Review

What is the value of each digit in 4,321?

4,321

4 = 4,000

3 = 300

2 = 20

1 = 1

Tip: Remember that each digit in a number is 10 times the place value of the digit at the right.

Write the value of the digits in each number as shown in the example above.

1. 8,935 ______________________________

2. 632 ______________________________

3. 1,761 ______________________________

4. 472 ______________________________

Look at the underlined digits in each number below. Write the place value of each. Then, tell how much greater the digit on the left is than the digit on the right.

5. <u>4,4</u>66; ______________________________

6. 6<u>77</u>; ______________________________

7. 8,<u>66</u>1; ______________________________

8. <u>2,2</u>90; ______________________________

Understand the Structure of Multi-Digit Numbers

Name ______________________________

Analyze numbers using place value.

1. Use your knowledge of place-value to tell how the underlined digits compare.

98,345 and 19,026

23,670 and 37,945

68,431 and 29,041

2. The Blue Whale's tongue weighs 5,952 pounds. A student says that the 5's in this number are 20 apart from each other. Is this correct or incorrect? Explain your answer using your knowledge of place value.

Read and Write Numbers to One Million

Name ______________________________

Review

A place value chart can help when reading and writing numbers.

Millions Period			Thousands Period			Ones Period		
hundreds	tens	ones	hundreds	tens	ones	hundreds	tens	ones
		3	5	4	6	1	8	9

Write the expanded form of the digits in 1,546,389.

1. 1 ______________

2. 6 ______________

3. 9 ______________

4. 4 ______________

Write these numbers in word form.

5. 1,522,100

6. 678,000

Write this number in standard form.

7. Nine hundred eighty-eight thousand, three hundred sixty-one

____

Read and Write Numbers to One Million

Name ______________________________

Reading and Writing World Populations

1. These numbers show approximate populations of various countries. Draw a line to its match.

Bhutan: 763,092	600,000 + 10,000 + 5,000 + 700 + 200 + 9
Solomon Islands: 669,823	three hundred thirty-nine thousand, thirty-one
Western Sahara: 582,463	600,000 + 60,000 + 9,000 + 800 + 20 + 3
Luxembourg: 615,729	two hundred ninety-nine thousand, eight hundred eighty-two
Iceland: 339,031	500,000 + 80,000 + 2,000 + 400 + 60 + 3
Vanuatu: 299,882	seven hundred sixty-three thousand, ninety-two

2. The approximate population of Cyprus is 1,198,575. Write this number in standard form, word form and expanded form.

Compare Multi-Digit Numbers

Name ______________________________

Review

Use a place-value chart to help you compare multi-digit numbers.

40,389 and 40,590

TIP: Compare each digit from left to right until you reach one that is larger. That is the greater number!

hundred thousands	ten thousands	thousands	hundreds	tens	ones
	4	0	3	8	9
	4	0	5	9	0

Which number is greater? Use a place-value chart to help you compare.

1. 30,989 or 30,777 __________

2. 66,688 or 66,639 __________

3. 90,991 or 90,999 __________

4. 100,001 or 100,009 __________

Compare each pair of numbers using the symbols <, >, or =.

5. 77,890 ◯ 77,340

6. 68,892 ◯ 68,899

7. 922,781 ◯ 923,783

8. 174,223 ◯ 172,301

9. 82,531 ◯ 82,431

10. 22,890 ◯ 21,999

11. 123,456 ◯ 123,654

12. 398,450 ◯ 398,430

Compare Multi-Digit Numbers

Name ______________________________

1. Compare these numbers. Then arrange them from least to greatest.

Numbers	**Least to Greatest**
88,980	**1.**
188,980	**2.**
1,888,990	**3.**
188,990	**4.**
88,990	**5.**
8,990	**6.**
1,888,980	**7.**
8,980	**8.**

2. Pick the two greatest numbers from the list. Compare them in two ways using the symbols <, >, or =.

Lesson **2-4 • Reinforce Understanding**

Round Multi-Digit Numbers

Name ____________________

Review

Use a number line to help you round numbers. Round 78,289 to the nearest thousand.

78,289 is closer to 78,000 than to 79,000.

78,289 rounded to the nearest thousand is 78,000.

Use a number line to help you round each number.

1. Round 40,189 to the nearest hundred. __________
2. Round 64,688 to the nearest thousand. __________
3. Round 80,791 to the nearest ten. __________
4. Round 123,001 to the nearest ten-thousand. __________

Which rounding is the better estimate?

5. A school wants to have enough cookies for a parent event. Last year 345 parents came. One teacher said they should buy 300 cookies. Another teacher said they should buy 400 cookies.

   Should the school round up or down? Explain your answer.

Round Multi-Digit Numbers

Name ______________________________

Rounding in the Real World

Think about each situation. Decide if it is a good idea to round the number or if an exact number would be better. Explain your answer for each.

1. A contractor has calculated that he needs 365 bags of cement to finish a job. He doesn't want to have any extra cement after the job is complete. How many should he purchase?

2. A city needs to decide how many medals to order for a community fun run. Last year, 1,937 people participated in the race. How many should the city leaders order?

Think and Solve.

You want to go to a movie with your cousin. You remember the movie costing about $7.45 last time you went. Would you round down to $7 or round up to $8? Explain your reasoning?

Estimate Sums or Differences

Name ____________________

Review

Use estimation strategies to check if your exact answer is reasonable.

A town has $4,000. Do they have enough to buy fireworks for $1,253 and cotton candy for $2,898?

Round to the nearest hundred.	Use front-end estimation.
1,253 + 2,898 = ?	1,253 + 2,898 = ?
Round to nearest hundred	Front-end estimation
1,300 + 2,900 = 4,200	1,000 + 2,000 = 3,000

The actual cost is $4,151. Rounding to the nearest hundred is the most reasonable estimation.

Solve the problem.

1. Tommy tracks how long he practices his music. He practices piano for 760 minutes and guitar for 230 minutes. How many minutes does he practice in all?

a. Round to the nearest hundred. __________

b. Use front-end estimation. __________

c. Find the **sum.** __________

d. Which method was closer to the actual sum?

____

2. A florist uses 8,453 flowers in June. She uses 3,598 flowers in July. How many flowers does she use during those two months?

a. Round to the nearest hundred. __________

b. Use front-end estimation. __________

c. Find the **sum.** __________

d. Which method was closer to the actual sum?

____

Estimate Sums or Differences

Name ______________________________

Estimating in the Real World

1. A family has saved this much money.

They have two wishes: buy tickets to an amusement park for $348 and buy some new sports equipment for $128. Can they make their wishes come true? Use rounding or front-end estimation to estimate. Explain your answer.

2. A student has this much money.

She owes a friend $33.00. After she pays her friend, will she have enough left over to buy a movie ticket for $8.00? Use rounding or front-end estimation to estimate. Explain your answer.

Strategies to Add Multi-Digit Numbers

Name ______________________________

Review

Use partial sums to help you add multi-digit numbers.

2,365 + 7,918 = ?

	2,365
	+ 7,918
5 + 8 =	**13**
60 + 10 =	**70**
300 + 900 =	**1,200**
2,000 + 7,000 =	**9,000**
	10,283

Add the partial sums to find the total.

2,365 + 7,918 = 10,283

Add 3,218 + 5,341 using partial sums.

1. Add the ones: 8 + ________ = ________
2. Add the tens: ________ + ________ = ________
3. Add the hundreds: ________ + 300 = ________
4. Add the thousands: ________ + ________ = ________
5. 3,218 + 5,341 = ________

Add.

6. 1,453 + 2,598 ________

Lesson **3–2 • Extend Thinking**

Strategies to Add Multi-Digit Numbers

Name ________________________________

1. The chart shows the weights in pounds of various objects. Add the numbers in each row. Use either strategy from the lesson.

Object	Weight
limousine	7,352 lbs
car	3,221 lbs
bell	2,080 lbs
truck	16,001 lbs
motorcycle	441 lbs

Find the total weights.

1. limousine and bell ____________

2. car and truck ____________

3. limousine and car ____________

4. limousine and motorcycle ____________

5. truck and motorcycle ____________

6. truck and limousine ____________

7. motorcycle and car ____________

8. Arrange the sums above in order from least to greatest.

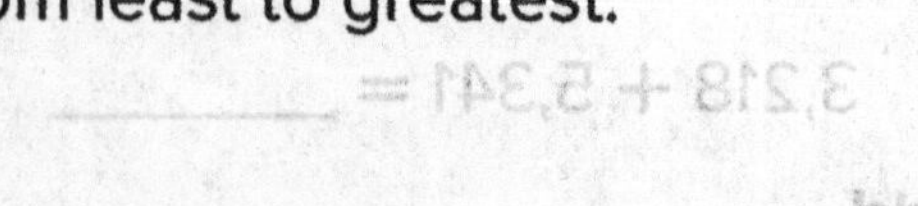

Lesson **3–3 • Reinforce Understanding**

Understand an Addition Algorithm

Name ____________________

Review

Use a place-value chart to set up the addition algorithm.

12,405 + 491 = ?

	ten thousands	thousands	hundreds	tens	ones
	1	2	4	0	5
+			4	9	1
	1	2	8	9	6

Start by adding the ones digits first.

12,405 + 491 = 12,896.

Add using the place value chart.

1. 2,408
 + 361

	thousands	hundreds	tens	ones
+				

2. 56,435
 + 3,054

	ten thousands	thousands	hundreds	tens	ones
+					

3. 6,384
 + 41,502

	ten thousands	thousands	hundreds	tens	ones
+					

Understand an Addition Algorithm

Name ______________________________

The tables show the number of people that visited a theme park on each Friday, Saturday, and Sunday of July.

	First Weekend	Second Weekend	Third Weekend	Fourth Weekend
Friday	14,328	10,413	12,163	11,422
Saturday	210	3,221	314	7,241
Sunday	5,101	6,024	7,221	1,016

1. Identify the weekend with the greatest number of visitors and the least number of visitors.

2. The table for the 2nd Weekend is incorrect. The tens digit for Saturday should be 4. How does this affect the order of the number of visitors from least to greatest over the four weekends in July?

Lesson **3–4 • Reinforce Understanding**

Understand an Addition Algorithm Involving Regrouping

Name ______________________________

Review

Use a place-value chart to set up the addition algorithm.

48,712 + 23,483 = ?

	ten thousands	thousands	hundreds	tens	ones
	1	1			
	4	8	7	1	2
+	2	3	4	8	3
	7	2	1	9	5

Start by adding the digits in the ones place. Regroup as needed.

48,712 + 23,483 = 72,195

Add using the place value chart.

1. 6,598
+ 2,361

	thousands	hundreds	tens	ones
+				

2. 47,942
+ 38,243

	ten thousands	thousands	hundreds	tens	ones
+					

Understand an Addition Algorithm Involving Regrouping

Name ______________________________

Jackson plays a video game 4 times. The number of points that Jackson earns for each level is shown below. He wins the video game when the total number of points is over 100,000.

	Level 1	Level 2	Level 3
First Time	28,456	36,211	41,048
Second Time	25,282	34,305	39,344
Third Time	26,336	37,498	40,553
Fourth Time	29,062	33,425	38,182

1. How many times did he win the video game? Which times did he win?

2. Without finding the exact sum, how can you determine which times Jackson won the game? Explain your answer and show your work.

Strategies to Subtract Multi-Digit Numbers

Name ______________________________

Review

Use place value to subtract.

Subtract 5,246 – 3,532 = ?

5,246 – 3,000 = 2,246

2,246 – 500 = 1,746

1,746 – 30 = 1,716

1,716 – 2 = 1,714

The difference is 1,714.

Use place value to subtract.

1. 3,455 – 1,246 = ________

3,455 – 1,000 = ________

________ – 200 = 2,255

________ – ________ = 2,215

2,215 – ________ = ________

2. 6,204 – 3,142 = ________

Lesson **3–5 • Extend Thinking**

Strategies to Subtract Multi-Digit Numbers

Name ______________________________

Skateboards on the Go wants to manufacture 5,462 skateboards by the end of the school year. Each day they manufacture 350 skateboards. After 5 days of manufacturing, how many skateboards will still need to be manufactured to reach this goal?

1. Use repeated subtraction and place value to determine the answer. Show your work.
After 5 days, the company will still need to manufacture ________ skateboard to reach their goal.

2. When will they reach their goal?

Understand a Subtraction Algorithm

Name ______________________________

Review

Use a place value chart to set up the subtraction algorithm.

68,694 − 7,504 = ?

	ten thousands	thousands	hundreds	tens	ones
	6	8	6	9	4
−		7	5	0	4
	6	1	1	9	0

Subtract the digits in the ones place first.

68,694 − 7,504 = 61,190

Subtract using the place value chart.

1. 8,656
− 245

	thousands	hundreds	tens	ones
−				

2. 49,875
− 6,324

	ten thousands	thousands	hundreds	tens	ones
−					

3. 92,784
− 81,612

	ten thousands	thousands	hundreds	tens	ones
−					

Understand a Subtraction Algorithm

Name ______________________________

For each problem, use the standard algorithm for subtraction to determine elapsed time.

Jamal's family started driving to his brother's college at 1:23 pm. They stopped at a rest stop at 4:36 pm and started driving again at 4:47 pm. They arrived at his brother's college at 7:59 pm.

1. How long did Jamal's family drive before they stopped at the rest stop? ______________________

2. How long were they at the rest stop? ______________________

3. How long did they drive after the rest stop before arriving at his brother's college? ______________________

4. How long was the trip? ______________________

5. Last year Jamal ran a marathon in 5 hours, 48 minutes, and 36 seconds or 5:48:36. Today he just finished a marathon in 4 hours, 25 minutes, and 2 seconds. How much faster did he run the marathon today than last year?

____

Understand a Subtraction Algorithm Involving Regrouping

Name ______________________________

Review

Use a place-value chart to set up the subtraction algorithm.

75,492 − 36,514 = ?

	ten thousands	thousands	hundreds	tens	ones
	6	14	14	8	12
	7	5	4	9	2
−	3	6	5	1	4
	3	8	9	7	8

Begin by subtracting the digits in the ones place. Regroup as needed.

75,492 − 36,514 = 38,978

Subtract using the place value chart.

1. 8,722 − 7,095 = ________

	thousands	hundreds	tens	ones
−				

2. 35,603 − 29,582 = ________

	ten thousands	thousands	hundreds	tens	ones
−					

Lesson **3–7 • Extend Thinking**

Understand a Subtraction Algorithm Involving Regrouping

Name ______________________________

The table shows the number of Earth days it takes a planet to orbit the sun.

Planet	Orbital Period in whole Earth days
Earth	365
Jupiter	4,333
Saturn	10,756
Uranus	30,687
Neptune	60,190

1. Use the subtraction algorithm and repeated subtraction to determine approximately how many times greater Uranus' orbital period is compared to Jupiter's orbital period. Show your work.

Uranus' orbital period is about ________ times greater than Jupiter's orbital period.

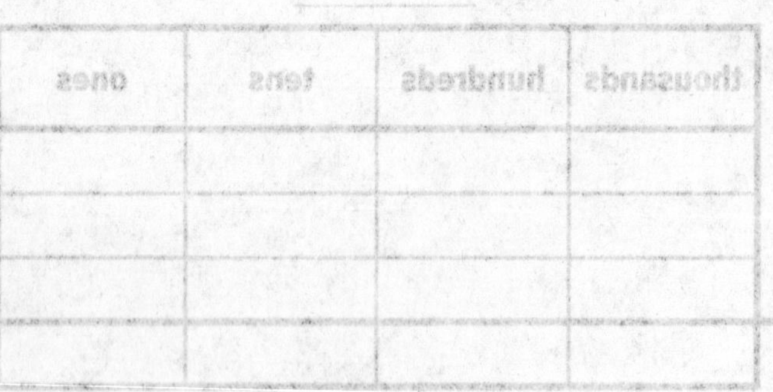

2. How much greater is Neptune's orbital period than Earth's? Show your work.

Represent and Solve Multi-Step Problems

Name ____________________

Review

In a stadium there are 8,749 red seats and 7,210 blue seats.

- **To change the design, 1,006 additional seats are made blue.**
- **Then 685 seats have the red cover removed.**

How many more seats are covered in blue than red now?

Step 1: Find the number of blue seats.

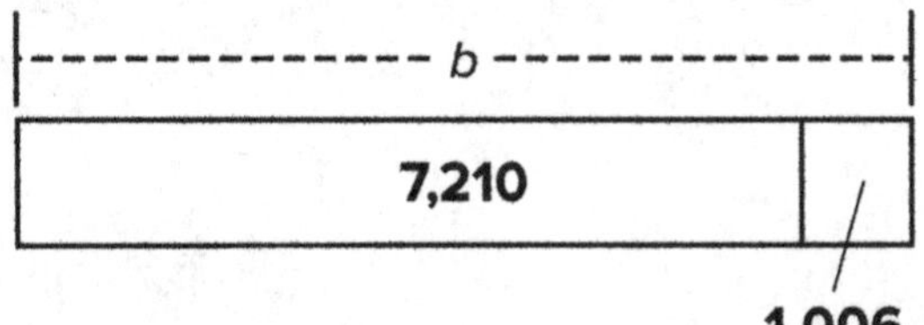

$7{,}210 + 1{,}006 = b$

$8{,}216 = b$

There are 8,216 blue seats.

Step 2: Find the number of red seats.

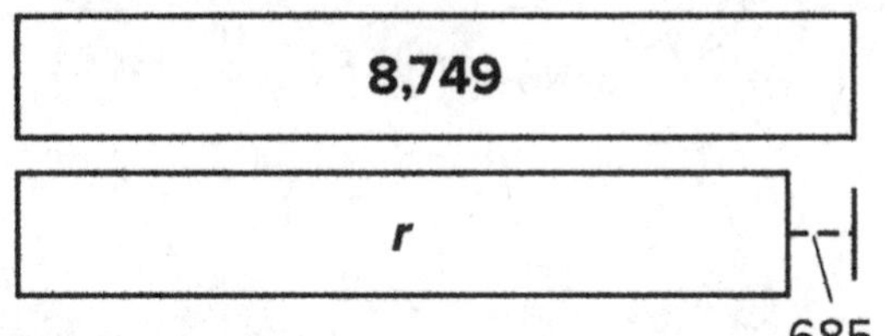

$r = 8{,}749 - 685$

$r = 8{,}064$

There are 8,064 red seats.

Step 3: Find how many more seats are blue than red now.

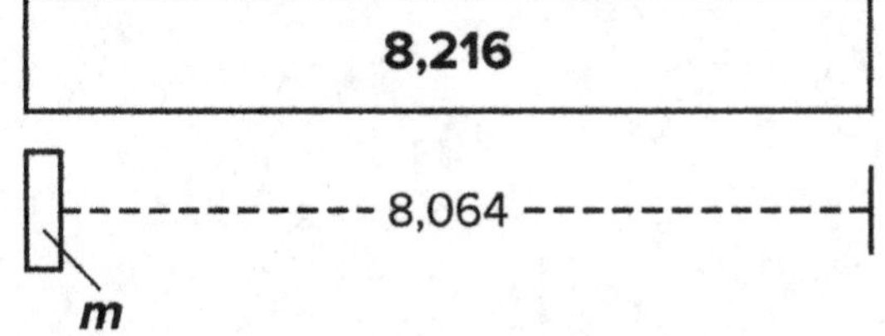

$m = 8{,}216 - 8{,}064$

$m = 152$

There are 152 more blue seats than red seats.

Use diagrams and equations to solve. Show your work.

1. A public library has 135,278 fiction books and 41,599 non-fiction books. Last week, 48,266 fiction books were checked out and 5,003 non-fiction books were checked out. How many books are in the library now? ____________________

Lesson **3–8 • Extend Thinking**

Represent and Solve Multi-Step Problems

Name

Write a word problem that takes at least two steps to solve. The beginning and ending numbers have been given to you. Then use a diagram and an equation to solve the problem. Show your work.

1. **Beginning:** 13,758

   Ending: 24,920

2. **Beginning:** 152,006

   Ending: 117,804

3. **Beginning:** 603,070

   Ending: 2,419

Solve Multi-Step Problems Involving Addition and Subtraction

Name ______________________________

Review

Some problems require more than one step to solve.

The current population of butterflies in Region A is 8,992. During the migration, 17,478 more butterflies appear. In Region B, there are 22,023 butterflies after the migration. How many more butterflies are in Region A than in Region B after the migration?

Step 1 Find how many butterflies are in Region A.

$17{,}478 + 8{,}992 = 26{,}470$

There are 26,470 butterflies in Region A.

Step 2 Find how many more butterflies are in Region A.

$26{,}470 - 22{,}023 = 4{,}447$

There are 4,447 more butterflies in Region A than in Region B.

Solve. Show your work.

1. The carnival sells tickets each day. Based on the data in the table, how many more tickets did they sell on Saturday and Sunday than on Wednesday, Thursday and Friday?

Day	Number of Tickets
Wednesday	8,159
Thursday	9,246
Friday	13,801
Saturday	24,632
Sunday	21,099

____

Solve Multi-Step Problems Involving Addition and Subtraction

Name ______________________________

A check register can help you keep track of how much money is in any financial account.

Here are the transactions for Julie's business account last month. The amounts in parentheses indicate withdrawals or money *subtracted* from the account. The other items are deposits or money *added* to the account.

- **Cash deposit** **$3,481**
- **Payroll** **($10,250)**
- **Taxes** **($2,105)**
- **Sales** **$24,982**
- **Customer Payment** **$9,010**
- **Rent** **($865)**
- **Supplies** **($1,499)**

Record all of the transactions in the check register and the current balance in the account. You can enter any date from January for each transaction. The first transaction has been recorded for you.

Date	Transaction	Deposit	Withdrawal	Balance
	Initial Balance			$9,276
01/05	Cash Deposit	3,481		$12,757
				$
				$
				$
				$
				$
				$

Understand Comparing with Multiplication

Name ______________________________

Review

You can use multiplicative comparison to relate quantities.

5 cubes 5 cubes 5 cubes 5 cubes 5 cubes 5 cubes

30 cubes

30 is 6 times as much as 5

$30 = 6 \times 5$

Solve the problem.

1. 48 is 8 times as much as 6

2. 35 is 7 times as much as 5

3. $21 = 3 \times 7$

4. $72 = 6 \times 12$

Write equations to solve the problem.

5. This weekend, Aiden sold 3 times as many granola bars as fruit rollups and 2 times as many fruit rollups as muffins. He sold 4 muffins. How many granola bars did Aiden sell? How many fruit rollups did Aiden sell? How many total baked items did Aiden sell this weekend?

Understand Comparing with Multiplication

Name ______________________________

Write a real-world situation involving multiplicative comparisons that may be represented by the equations shown.

1. $24 = 4 \times 6; 24 = 6 \times 4$

2. $45 = 9 \times 5; 45 = 5 \times 9$

3. $36 = 3 \times 12; 36 = 12 \times 3$

4. $60 = 12 \times 5; 60 = 5 \times 12$

Represent Comparison Problems

Name ______________________________

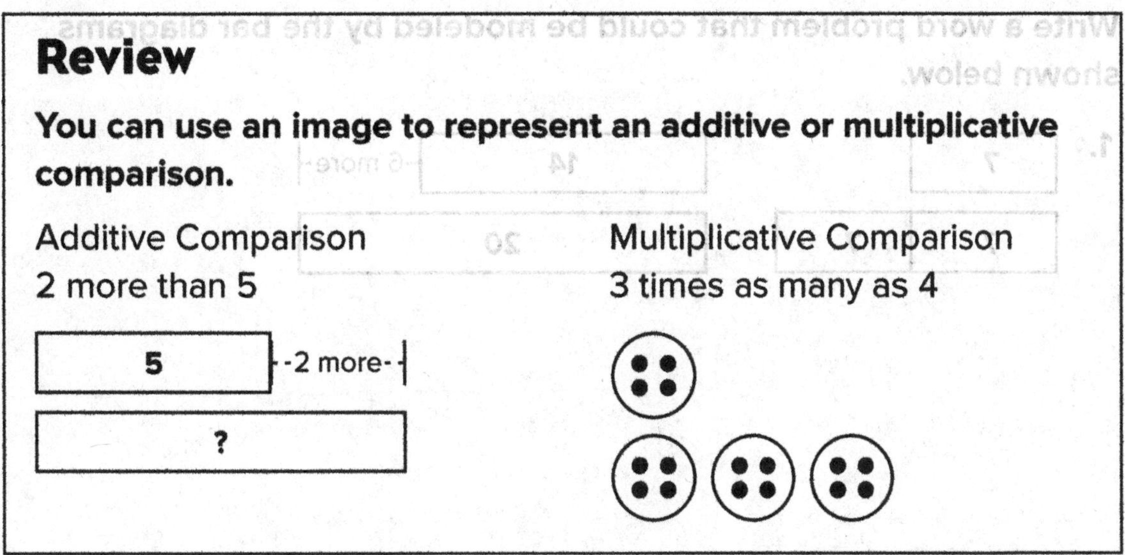

Review

You can use an image to represent an additive or multiplicative comparison.

Additive Comparison
2 more than 5

Multiplicative Comparison
3 times as many as 4

Write an equation to represent the comparison. Then solve.

1. 7 more than 9 ______________
2. 6 times as long as 2 cm ______________
3. 8 more than 11 ______________
4. 4 times as many as 8 ______________

Use bar diagrams and equations to solve.

5. Brad baked 36 muffins. Adam baked 8 more muffins than Brad. Fred baked 2 times as many muffins as Adam. How many muffins did Adam bake? How many muffins did Fred bake?

Represent Comparison Problems

Name ______________________________

Write a word problem that could be modeled by the bar diagrams shown below.

1.

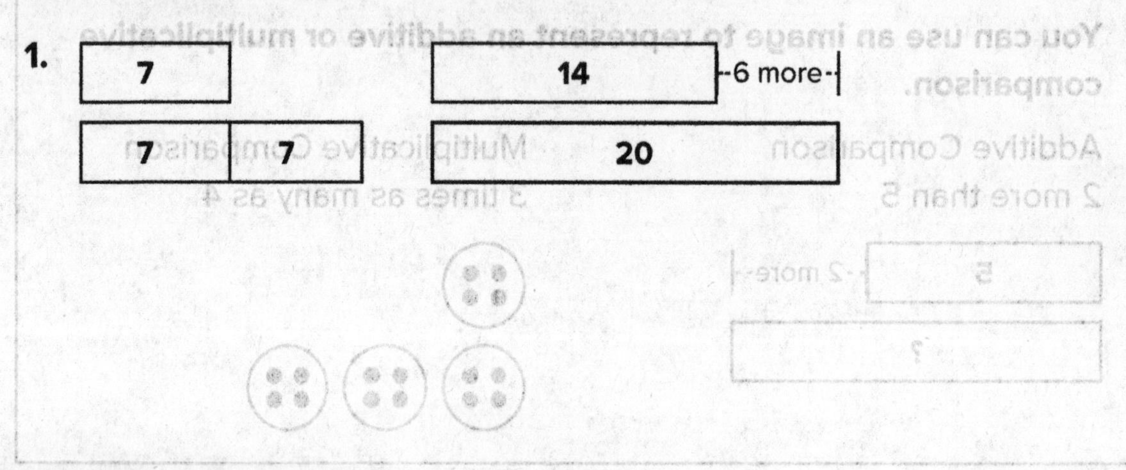

2.

4
8 more
12
12
12
12

3.

9
4 more
13
13
13
13
13

Solve Comparison Problems Using Multiplication

Name ______________________________

Review

You can use a bar diagram and equation to represent a multiplicative comparison involving an unknown factor.

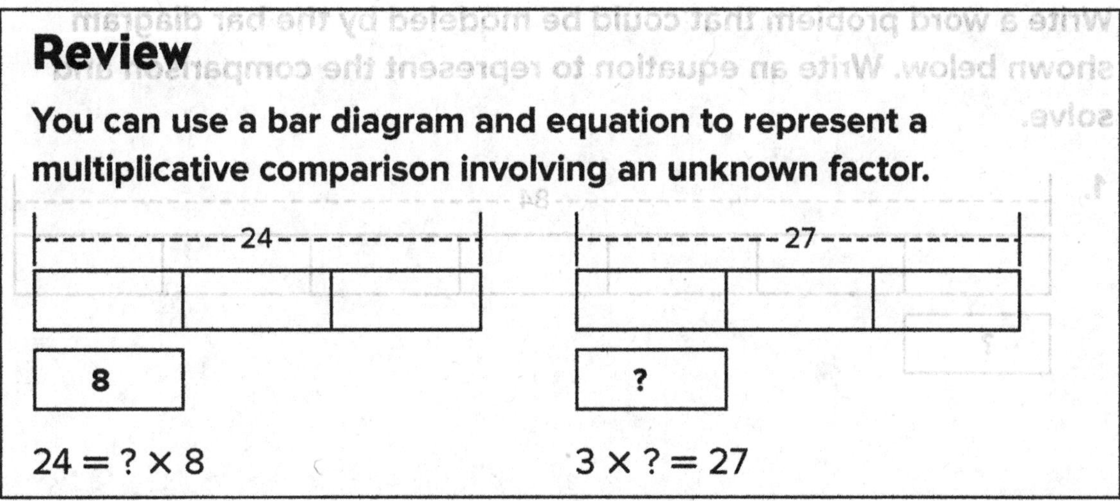

Write an equation to represent the comparison. Then solve.

1. 63 is 7 times as much as ?

   ____

2. 88 is ? times as many as 11

   ____

3. 48 is ? times as much as 12

   ____

4. 28 is 4 times as many as ?

   ____

Write an equation and solve. Use a bar diagram, if needed.

5. Mara planted 35 rose bushes. She planted 7 times as many rose bushes as peach trees. How many peach trees did she plant?

6. Carly sold 56 raffle tickets this week. How many times more tickets did she sell this week than last week, if she sold 8 tickets last week?

Lesson **4-3 • Extend Thinking**

Solve Comparison Problems Using Multiplication

Name ______________________________

Write a word problem that could be modeled by the bar diagram shown below. Write an equation to represent the comparison and solve.

1.

84

?

2.

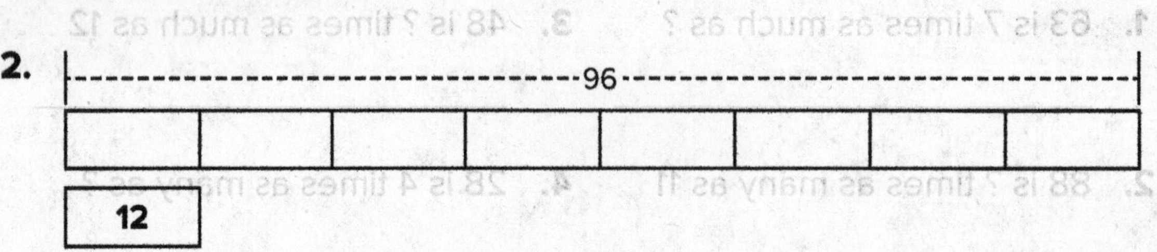

3.

Solve Comparison Problems Using Division

Name ______________________________

Review

You can use a bar diagram and a division equation to represent a multiplicative comparison.

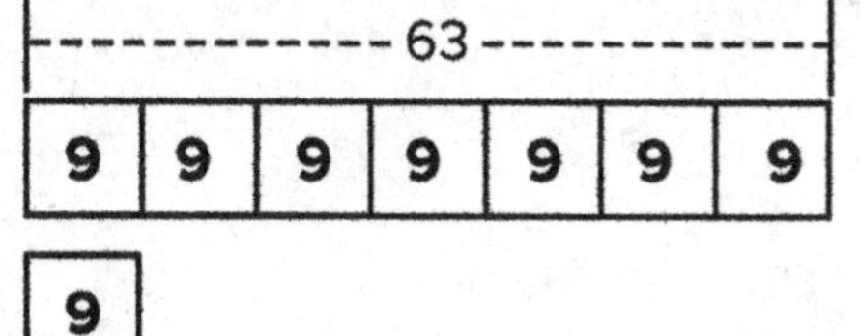

45

?

$63 \div 9 = ?$

$7 = ?$

$45 \div 5 = ?$

$9 = ?$

Write and solve a division equation to represent the comparison.

1. 99 is 11 times as much as ?

____

2. 3 times ? is 24

____

3. 42 in. is 7 times as long as ?

____

4. 16 oz. is 8 times as much as ?

____

Write an equation and solve. Use a bar diagram, if needed.

5. Penelope hit 15 home runs this season. Kori hit 5 home runs. How many times as many home runs did Penelope hit as Kori?

6. Kim sold 48 boxes of crackers. This is 8 times as many boxes as Robert sold. How many boxes did Robert sell?

Lesson **4-4 • Extend Thinking**

Solve Comparison Problems Using Division

Name ______________________________

Write a word problem that could be modeled by the bar diagram shown below. Write a division equation to represent the comparison and solve.

1.

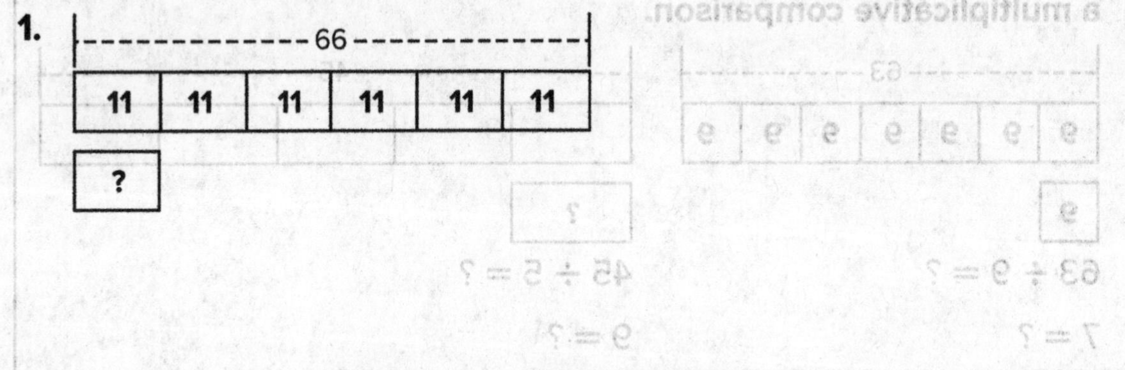

2.

36

?

3.

84

7 7 7 7 7 7 7 7 7 7 7 7

7

Understand Factors of a Number

Name ______________________________

Review

You can use arrays to find factor pairs of a number.

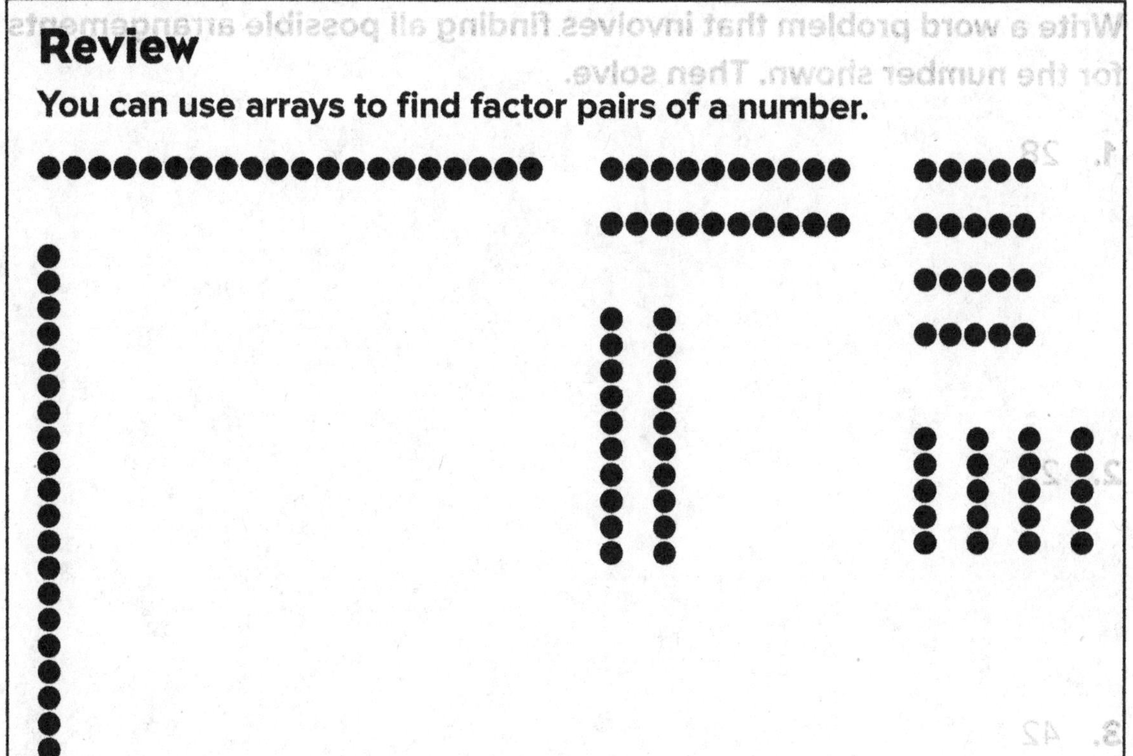

The factor pairs of 20 are 1 and 20, 2 and 10, and 4 and 5.

Draw a model to find all factor pairs of each number.

1. 16

2. 19

Draw a model to find all the possible arrangements.

3. A bakery has 36 pastries to arrange on its shelves. How can the pastries be arranged if the owner wishes to put an equal number of pastries on each shelf and arrange them on 2 to 6 shelves?

Understand Factors of a Number

Name ____________________

Write a word problem that involves finding all possible arrangements for the number shown. Then solve.

1. 28

2. 25

3. 42

4. 36

Understand Prime and Composite Numbers

Name ______________________________

Review

You can use arrays to show whether a number is prime or composite.

5 × 1	6 × 1	2 × 3
1 × 5	1 × 6	3 × 2

The number, 5, is prime because it only has one factor pair: 1 and 5.
The number, 6, is composite because it has more than 1 factor pair.

State whether each number is prime or composite. Explain your thinking.

1. 3 ______________________________

____

2. 16 ______________________________

____

3. 42 ______________________________

____

____

4. Find a prime number between 90 and 100. How do you know it is prime?

____

____

Understand Prime and Composite Numbers

Name ______________________________

State whether each number is prime or composite. Write a real-world situation that involves the arrangement of items in one or more equal groups for each number given. Explain how the situation shows that the number is prime or composite.

1. 13

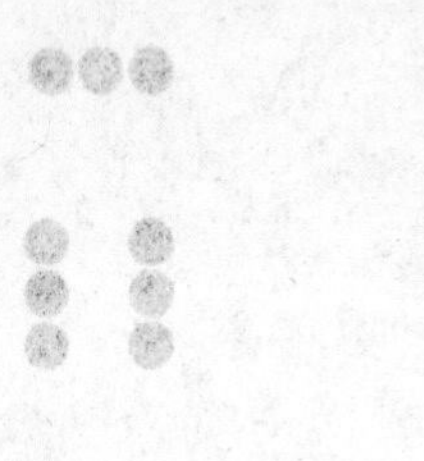

2. 51

3. 43

4. 94

Understand Multiples

Name ______________________________

Review

You can use a number line to show multiples of a number.

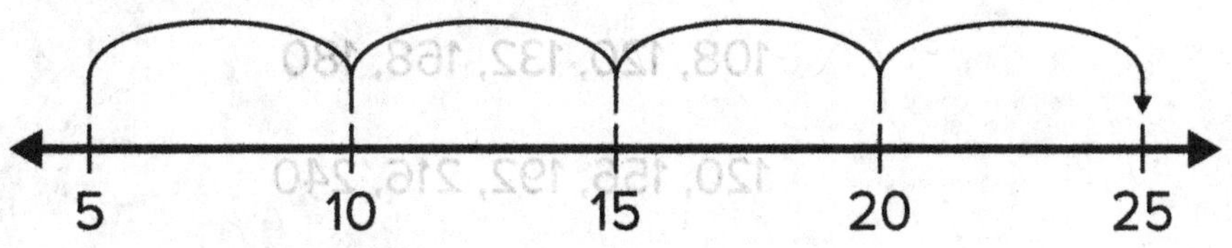

5, 10, 15, 20, and 25 are multiples of 5.

Find 5 multiples of each number.

1. 4

2. 9

3. 7

4. 12

Solve using multiples. Show your work.

5. Each book is worth 6 points. Hannah will read 3 to 5 books. How many points might Hannah earn?

6. Each letter uses 2 stamps. Mike needs to mail 6 to 8 letters. How many stamps might Mike use?

7. Each photo costs 3 dollars. Lindsey will take 2 to 5 photos. How many dollars might Lindsey spend on photos?

Understand Multiples

Name ______________________________

Match each number to one set of multiples.

1.	4	112, 128, 152, 176, 192
2.	7	108, 120, 132, 168, 180
3.	9	120, 156, 192, 216, 240
4.	8	160, 200, 240, 300, 340
5.	12	104, 116, 136, 152, 164
6.	10	112, 154, 182, 224, 252
7.	6	112, 126, 147, 161, 182
8.	14	108, 144, 180, 216, 252
9.	5	132, 198, 242, 264, 308
10.	20	126, 162, 180, 216, 234
11.	22	120, 150, 170, 190, 200
12.	18	100, 130, 165, 180, 210

Explain how you matched the numbers to the multiples.

Lesson **5-4 • Reinforce Understanding**

Number or Shape Patterns

Name ______________________________

Review

There are number patterns and shape patterns.

7, 9, 11, 13, 15

This is a number pattern that is described by the rule add 2.

This is a shape pattern that grows. It may be described by the rule: 1 triangle, 1 rectangle, 1 diamond, repeat, but increase number of diamonds by 1.

State the rule for each pattern.

1. 10, 13, 16, 19 __________

2. 4, 8, 16, 32 ____________________

3. 6, 9, 12, 6, 9, 12 ____________________

4.

5.

Write the next three terms in the sequence. Explain your thinking.

6. 45, 40, 35, 30, _____, _____, _____

Number or Shape Patterns

Name ______________________________

Draw a repeating or growing pattern, as indicated.

1. number pattern that repeats

2. number pattern that grows by adding a constant

3. number pattern that grows by multiplying by a factor

4. shape pattern that repeats

5. shape pattern that grows

Write a word problem that involves a number pattern for the given rule. Use the pattern to solve. Show your work.

6. Add 4

Lesson **5-5 • Reinforce Understanding**

Generate a Pattern

Name ______________________________

Review

Use a table to find a later term of a pattern.

Starting with 4, add 6. What is the 9th term of the pattern?

Order in Pattern	Equation	Term in Pattern
1st	$(1 - 1) \times 6 + 4$	4
2nd	$(2 - 1) \times 6 + 4$	10
3rd	$(3 - 1) \times 6 + 4$	16
9th	$(9 - 1) \times 6 + 4$	52

The 9th term of the pattern is 58.

Write or draw the first five terms of each pattern.

1. Starting with 12, add 7 ______________________

2. Starting with 20, subtract 3 ______________________

3. 1 AB, 2 C's repeating ______________________

4. 2 W's, XY, 2 Z's repeating ______________________

What is the 10th term of each pattern?

5. 50, 55, 60, 65, ···

____

____

6. 7, 15, 23, 31, ···

____

____

7. Malcolm drives 5 miles to the pool and back each day for swim practice. How many miles does he drive in 5 days? In 10 days

Generate a Pattern

Name ______________________________

Use the table to find the first four terms and the 12th term for each pattern.

1. Starting with 8, add 8

Order in Pattern	Equation	Term in Pattern
1st	1×8	8
2nd		
3rd		
4th		
12th		

2. Starting with 12, add 10

Order in Pattern	Equation	Term in Pattern
1st	$(1 - 1) \times 10 + 12$	12
2nd		
3rd		
4th		
12th		

3. Compare the equations for each pattern in problems 1 and 2. What is similar? What is different?

4. Write an equation for each pattern in problems 1 and 2 for an unknown *nth* term. ______________________

Analyze Features of a Pattern

Name ______________________________

Review

Use a table to analyze a pattern to find features that are not stated in the pattern rule. Start with 3, add 5.

Order in the Pattern	1st	2nd	3rd	4th	5th
Term	3	8	13	18	23

The terms alternate odd then even.

The ending digits of the terms alternate 3 then 8.

What is a feature of each pattern that is not stated in the pattern rule? Explain why this feature exists.

1. Start with 7, add 14.

2. Start with 9, add 2.

3. Start with 6, add 8.

4. Start with 11, add 9.

Analyze Features of a Pattern

Name

1. Create a pattern that starts with an even number and subtracts an even number. What is a feature of the pattern that is not stated in the pattern rule? Explain why this feature exists.

2. Micah creates a square with side length 4 inches. He increases the length of the side of each square by 1 inch as he makes each. If he continues this pattern, what would be the area of the 6th square?

3. Create a pattern that starts with an odd number and subtracts an odd number. What is a feature of the pattern that is not stated in the pattern rule? Explain why this feature exists.

4. Create a pattern that starts with an odd number and subtracts an even number. What is a feature of the pattern that is not stated in the pattern rule? Explain why this feature exists.

Multiply by Multiples of 10, 100, or 1,000

Name ______________________________

Review

There are different ways to multiply a factor by a multiple of 10.

Use multiplication facts and place value	**Use the Associative Property of Multiplication**
$n = 5 \times 700$	$n = 400 \times 6$
$n = 5 \times 7$ hundreds	$n = 4 \times 100 \times 6$
$n = 35$ hundreds	$n = 4 \times 6 \times 100$
$n = 3{,}500$	$n = 24 \times 100$
	$n = 2{,}400$

What's the product? Complete the equation.

1. $3 \times 200 = 3 \times$ __________ hundreds

= __________ hundreds

= __________

2. $4 \times 9{,}000 = 4 \times$ __________ thousands

= __________ thousands

= __________

Solve each word problem.

3. Judy has 8 bags of marbles. Each bag contains 20 marbles. How many marbles does Judy have in all?

4. Victor buys 7 packages of paper. Each package contains 500 sheets of paper. How many sheets will he have in all?

Multiply by Multiples of 10, 100, or 1,000

Name ______________________________

Write a word problem that involves finding the product of the numbers given. Then solve.

1. 9 × 300

2. 6 × 50

3. 4 × 2,000

4. 3 × 600

5. 8 × 30

Estimate Products

Name ______________________________

Review

There are different ways to estimate products.

Use compatible numbers.	Use rounding.
157 × 2 = *n*	157 × 2 = *n*
150 × 2 = 300	200 × 2 = 400
about 300	about 400

How can you use compatible numbers to estimate the product? Complete the equation.

1. 269 × 5 = ?

   ________ × 5 = ________

2. 4 × 742 = ?

   4 × ________ = ________

How can you use rounding to estimate the product? Complete each equation.

3. 3 × 1,289 = ?

   3 × ________ = ________

4. 624 × 6 = ?

   ________ × 6 = ________

Use estimation to solve each word problem.

5. Each camera costs 114 dollars. About how much will 4 cameras cost?

   ____

6. Each day, a school cafeteria serves 1,119 meals. About how many meals does the cafeteria serve in 5 days?

   ____

Estimate Products

Name ______________________________

Find a possible value for *n* that makes each statement true. Explain your thinking.

1. The product of $4 \times n$ is about 8,000.

2. The product of $n \times 4$ is about 1,000.

3. The product of $6 \times n$ is about 24,000.

4. The product of $n \times 2$ is about 7,000.

5. The product of $8 \times n$ is about 40,000.

6. The product of $n \times 9$ is about 27,000.

Use the Distributive Property to Multiply

Name ______________________________

Review

You can decompose a factor and apply the Distributive Property of Multiplication to multiply numbers.

One Way	Another Way
$8 \times 14 = 8 \times (10 + 4)$ $= (8 \times 10) + (8 \times 4)$ $= 80 + 32$ $= 112$	$8 \times 14 = 8 \times (7 + 7)$ $= (8 \times 7) + (8 \times 7)$ $= 56 + 56$ $= 112$

How can you use the Distributive Property to find the product? Write and solve an equation to show your work.

1. $5 \times 8 = 5 \times (\_\_\_\_ + \_\_\_\_)$

   $= (5 \times \_\_\_\_) + (5 \times \_\_\_\_)$

   $= \_\_\_\_ + \_\_\_\_$

   $= \_\_\_\_$

2. $6 \times 17 = 5 \times (\_\_\_\_ + \_\_\_\_)$

   $= (6 \times \_\_\_\_) + (6 \times \_\_\_\_)$

   $= \_\_\_\_ + \_\_\_\_$

   $= \_\_\_\_$

3. Ann bakes 3 batches of muffins. Each batch has 24 muffins. How many muffins does Ann bake in all?

Use the Distributive Property to Multiply

Name ______________________________

Use each given product to write a multiplication equation. Then show how a factor can be decomposed in more than one way to find the product.

1. 48

2. 56

3. 60

4. 42

5. 72

6. 64

Multiply 2-Digit by 1-Digit Factors

Name ______________________________

Review

You can use an area model and the Distributive Property to find partial products used to calculate a product.

$a = 7 \times 36$

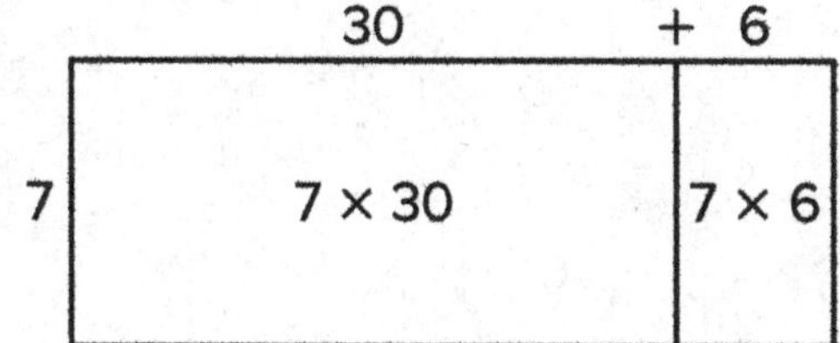

$a = (7 \times 30) + (7 \times 6)$

$a = 210 + 42$

$a = 252$

How can you decompose a factor and find the partial products? Use an area model to show your work.

1. $8 \times 47 =$

2. $92 \times 5 =$

3. $6 \times 56 =$

4. A bottle of lotion has 8 ounces of lotion. How much lotion is in 15 bottles?

Multiply 2-Digit by 1-Digit Factors

Name

Draw an area model that may be used to represent each multiplication problem. Use the area model and the Distributive Property of Multiplication to calculate the product.

1. 102 × 7

2. 3 × 806

3. 405 × 5

4. 9 × 307

5. 903 × 2

6. 6 × 408

Multiply Multi-Digit by 1-Digit Factors

Name ____________________

Review

An area model and the Distributive Property of Multiplication can be used to find partial products when multiplying a multi-digit factor by a 1-digit factor.

$b = 3 \times 246$

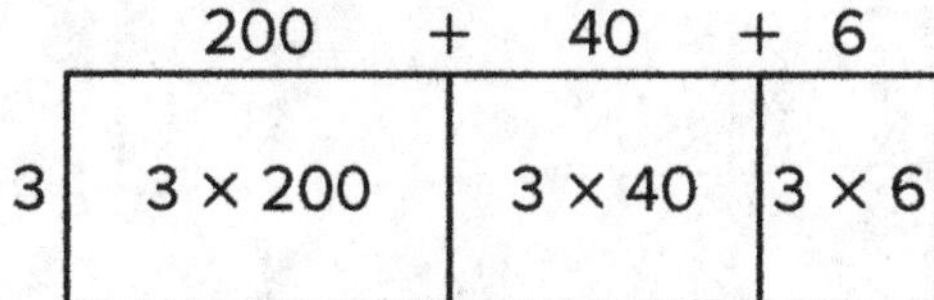

$b = (3 \times 200) + (3 \times 40) + (3 \times 6)$

$b = 600 + 120 + 18$

$b = 738$

How can you find the product? Use an area model and partial products to solve.

1. $8 \times 256 =$

2. $3{,}425 \times 5 =$

3. $8 \times 927 =$

4. $2{,}484 \times 9 =$

5. A bottle contains 591 mL of water. How much water is in 6 bottles?

Multiply Multi-Digit by 1-Digit Factors

Name ____

Use each given product to write a multiplication equation. Then use an area model and decomposition of a factor to represent the product.

1. 650

2. 1,100

3. 1,250

4. 1,500

5. 4,400

6. 3,600

Multiply Two Multiples of 10

Name ______________________________

Review

There is more than one way to multiply two multiples of 10. You can use multiplication facts and place value or decomposition and the Associative Property of Multiplication to multiply two numbers.

One Way	Another Way
$20 \times 40 = 20 \times 4$ tens $= 80$ tens $= 800$	$20 \times 40 = 2 \times 10 \times 4 \times 10$ $= 2 \times 4 \times 10 \times 10$ $= 8 \times 10 \times 10$ $= 80 \times 10$ $= 800$

How can you find the product? Complete the equation.

1. $80 \times 20 =$ ______________________________

2. $30 \times 40 =$ ______________________________

3. $60 \times 50 =$ ______________________________

4. $70 \times 90 =$ ______________________________

How can you solve this problem? Show your work.

5. Each box contains 30 nails. How many nails are in 80 boxes?

Multiply Two Multiples of 10

Name ______________________________

Write a word problem to represent each product. Then solve.

1. 90 × 30

2. 20 × 70

3. 50 × 50

4. 80 × 40

5. 60 × 30

Lesson **6-7 • Reinforce Understanding**

Multiply Two 2-Digit Factors

Name ______________________________

Review

You can use an area model and partial products to multiply two 2-digit factors.

$d = 38 \times 24$

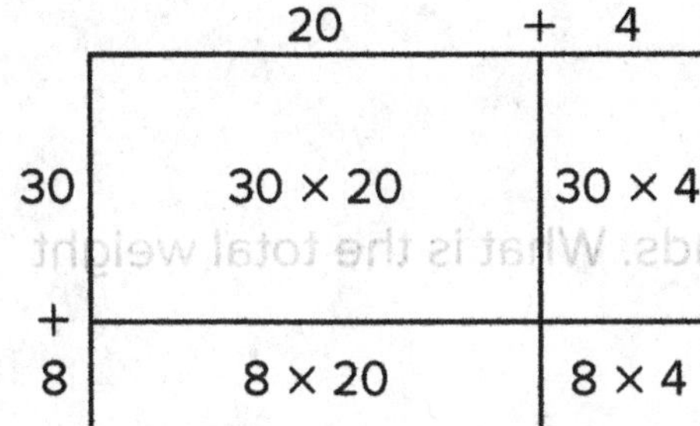

$d = (30 + 8) \times (20 + 4)$

$d = (30 \times 20) + (30 \times 4) + (8 \times 20) + (8 \times 4)$

$d = 600 + 120 + 160 + 32$

$d = 912$

How can you use partial products to solve? Use an area model to show your work.

1. $42 \times 25 =$

2. $34 \times 46 =$

3. $67 \times 92 =$

4. $26 \times 34 =$

Multiply Two 2-Digit Factors

Name ______________________________

Fill in two 2-digit numbers to complete each problem. Use an area model to represent the multiplication. Then solve.

1. Robert drives ____________ miles each day. How many miles does he drive in ____________ days?

2. Each box weighs ____________ pounds. What is the total weight of ____________ of the boxes?

3. Each day, Eric completes ____________ sit-ups. How many sit-ups does Eric complete in ____________ days?

4. A geologist has ____________ rock collections. Each collection contains ____________ rocks. How many rocks does the geologist have?

5. Each pack contains ____________ tomato plants. How many tomato plants are in ____________ packs?

Solve Multi-Step Problems Involving Multiplication

Name ______________________________

Review

You can solve a multi-step problem involving multiplication by writing out the steps.

A teacher buys sandwiches and fruit cups for her class. Each sandwich costs $4. Each fruit cup costs $2. How much will she spend if she buys 25 sandwiches and 22 fruit cups?

Step 1: Find the cost of the sandwiches.	$25 \times 4 = 4 \times 25$ $= 4 \times (20 + 5)$ $= (4 \times 20) + (4 \times 5)$ $= 80 + 20$ $= 100$
Step 2: Find the cost of the fruit cups.	$22 \times 2 = 2 \times 22$ $= 2 \times (20 + 2)$ $= (2 \times 20) + (2 \times 2)$ $= 40 + 4$ $= 44$
Step 3: Find the total cost.	$100 + 44 = 144$ The teacher will spend $144.

How can you represent the problem situation? Solve using partial products.

Kenny finishes 4 math assignments and 6 science assignments each week. Each math assignment has 12 problems. Each science assignment has 18 problems. How many total problems does Kenny finish each week?

Solve Multi-Step Problems Involving Multiplication

Name ______________________________

Given each problem, fill in two 2-digit numbers and then solve. Show your work.

Students and school staff purchase supplies from the school bookstore.

1. Marcus buys packages of pencils and packages of pens. Each package of pencils contains _____ pencils. Each package of pens contains _____ pens. Marcus buys _____ packages of pencils and _____ packages of pens. How many total pencils and pens does Marcus buy?

2. A teacher buys packages of construction paper and packages of notebook paper. Each package of construction paper contains _____ sheets. Each package of notebook paper contains _____ sheets. The teacher buys _____ packages of construction paper and _____ packages of notebook paper. How many total sheets of paper does the teacher buy?

Divide Multiples of 10, 100, and 1,000

Name ______________________________

Review

You can use place value to divide a multiple of 10, 100, and 1,000.

32 ÷ 4 = 8

32 tens ÷ 4 = 8 tens so 320 ÷ 4 = 80

32 hundreds ÷ 4 = 8 hundreds so 3,200 ÷ 4 = 800

32 thousands ÷ 4 = 8 thousands so 32,000 ÷ 4 = 8,000

Find the quotient. Complete each equation.

1. 63 ones ÷ 9 = ________ ones

63 tens ÷ 9 = ________ tens

63 ____________ ÷ 9 = 7 hundreds

2. 420 ÷ 7 = ________

4,200 ÷ 7 = ________

42,000 ÷ 7 = ________

3. 330 ÷ 11 = ________ tens ÷ 11 so, 330 ÷ 11 = ________

Solve each word problem.

4. Michelle has 280 pencil toppers to share equally with 7 friends. How many pencil toppers will each friend receive?

5. Carmen has 2,100 recipes. She plans to place an equal number of recipes in 7 containers. How many recipes will go in each container?

Divide Multiples of 10, 100, and 1,000

Name ________________________________

Match each equation on the left with the correct answer on the right. Place the equation number on the blank.

1.	640 ÷ 8 = ?	8 thousands ________
2.	28 ÷ 7 = ?	6 tens ________
3.	3,000 ÷ 5 = ?	4 thousands ________
4.	40 ÷ 5 = ?	8 hundreds ________
5.	3,600 ÷ 9 = ?	6 ones ________
6.	3,200 ÷ 4 = ?	6 thousands ________
7.	28,000 ÷ 7 = ?	4 tens ________
8.	24,000 ÷ 4 = ?	4 ones ________
9.	240 ÷ 6 = ?	8 tens ________
10.	24 ÷ 4 = ?	6 hundreds ________
11.	48,000 ÷ 6 = ?	4 hundreds ________
12.	360 ÷ 6 = ?	8 ones ________

Use division facts and place value OR the relationship between multiplication and division to solve for the variable. Show your work.

13. $450 \div a = 90$

14. $5{,}400 \div m = 60$

15. $84{,}000 \div 7 = c$

16. $7{,}200 \div 8 = p$

Estimate Quotients

Name ____________________

Review

You can use compatible numbers and a range to estimate a quotient.

Use a compatible number less than the dividend.	Use a compatible number greater than the dividend.
$341 \div 8 = ?$ ↓ $320 \div 8 = 40$	$341 \div 8 = ?$ ↓ $400 \div 8 = 50$

The estimated quotient is about 40 to 50.

Find the estimated quotient.

1. $728 \div 8$

2. $4{,}362 \div 7$

Use compatible numbers to find an estimated range for the quotient.

3. $532 \div 5$

4. $6{,}208 \div 9$

Solve each word problem.

5. A fitness tracker shows that Elizabeth made 3,450 steps in 9 hours. If she made about the same number of steps each hour, about how many steps did she make each hour?

6. A special event has 245 guests. The guests will be divided into 8 groups. If each group has about the same number of guests, about how guests will be in each group?

Estimate Quotients

Name

Find a possible value for *n* that makes each statement true. Explain your thinking.

1. The estimated quotient of 463 ÷ *n* is about 60.

2. The estimated quotient of 562 ÷ *n* is about 110.

3. The estimated quotient of 249 ÷ *n* is about 40.

4. The estimated quotient of 3,424 ÷ *n* is about 500.

5. The estimated quotient of 9,248 ÷ *n* is about 1,000.

6. The estimated quotient of 5,471 ÷ *n* is about 700.

Find Equal Shares

Name ______________________________

Review

You can use counters and groups to share equally.

54 ÷ 3 = ?

Place 54 counters into 3 groups until none are left.

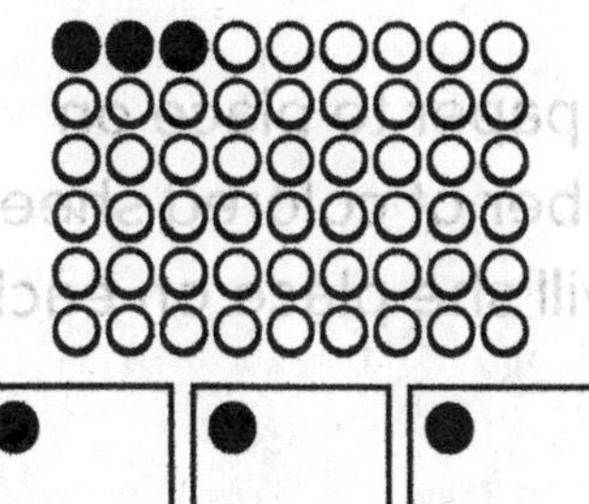

Count the number of counters in each group.

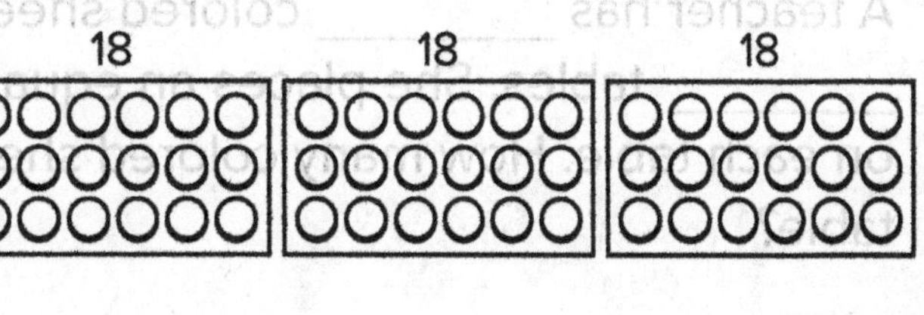

There are 18 counters in each group.

54 ÷ 3 = **18**

Solve.

1. 15 ÷ 5 = ____________

2. 42 ÷ 6 = ____________

3. 44 ÷ 4 = ____________

4. 84 ÷ 7 = ____________

Solve using counters or pictures.

5. Jenny has to solve 36 math problems. She plans to solve 4 problems each day. How many days will it take her to solve all of the math problems?

Find Equal Shares

Name ______________________

Fill in two numbers to complete each problem to show equal sharing. Use a picture to represent the problem. Then solve.

1. Kylie has ________ packages of crackers. She gives an equal number of packages to ________ friends. How many packages of crackers does each friend receive?

2. A teacher has ________ colored sheets of paper to place on ________ tables. She places an equal number of colored sheets on each table. How many colored sheets will she place on each table?

3. Michelle has earned ________ points for completing ________ math challenges. She earns the same number of points for each challenge. How many points does she earn for each challenge?

4. Kori needs to plant ________ pumpkin seeds. She will plant an equal number of seeds in ________ gardens. How many pumpkin seeds will she plant in each garden?

5. A principal plans to place photos of ________ award winners on the shelves in her office. She will place the same number of photos on each shelf. If the principal uses ________ shelves, how many photos will she place on each shelf?

Lesson **7-4 • Reinforce Understanding**

Understand Partial Quotients

Name ______________________________

Review

You can use base-ten blocks and partial quotients to divide.

$144 \div 6 = ?$

Subtract 6 groups of 20 from the total.

Subtract 6 groups of 4 from the total.

$20 + 4 = 24$, so $144 \div 6 = 24$

Use a representation and partial quotients to divide.

1. $130 \div 5 = ?$

2. $162 \div 9 = ?$

3. $216 \div 8 = ?$

4. $306 \div 9 = ?$

Understand Partial Quotients

Name ______________________________

Use partial quotients and a representation to solve.

Elementary school students will attend a county fair.

1. A local restaurant has a booth at the fair and is giving away 264 gift cards. The restaurant will give away the same number of gift cards each day for 4 days. How many gift cards will the restaurant give away each day?

2. A total of 112 students attended a magic show. The students were seated on 8 rows of bleachers with an equal number of students on each row. How many students were on each row of bleachers?

3. A parent spends 108 dollars on funnel cakes for all of the students in a class. Each funnel cake costs 6 dollars. How many students are in the class?

4. A student wants to save 135 dollars to donate to a charity that will be giving away items at the fair. The student plans to save 9 dollars each week. After how many weeks will Eric have enough money to make his donation?

Divide 4-Digit Dividends by 1-Digit Divisors

Name ______________________________

Review

You can use an area model and partial quotients to divide.

$c = 2{,}550 \div 6$

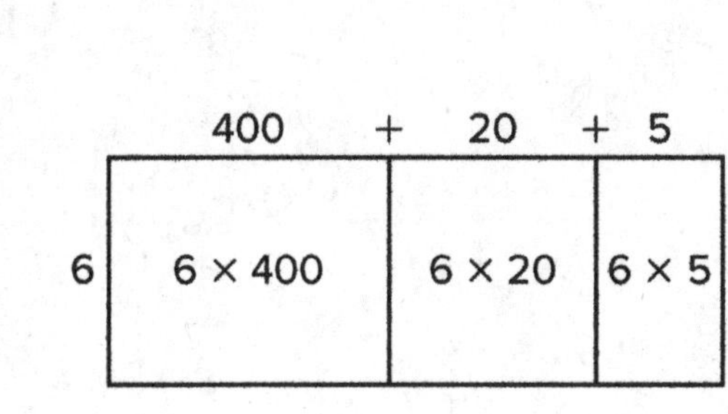

$$\begin{array}{r l} 2{,}550 & \\ -\underline{2{,}400} & (6 \times 400) \\ 150 & \\ -\underline{120} & (6 \times 20) \\ 30 & \\ -\underline{\;\;30} & (6 \times 5) \\ 0 & \end{array}$$

$400 + 20 + 25 = 425$

$c = 2{,}550 \div 6$

$c = 425$

Use an area model and partial quotients to divide.

1. $1{,}180 \div 5 =$ ____________

2. $1{,}788 \div 4 =$ ____________

Use partial quotients to solve.

3. A new company has 5,200 free samples to give away. The company will give away an equal number of samples each month for 8 months. What is the maximum number of samples that the company will give away each month?

Divide 4-Digit Dividends by 1-Digit Divisors

Name ______________________________

Write a word problem that may be solved by calculating the quotient. Then use an area model and partial quotients to solve.

1. 2,920 ÷ 8

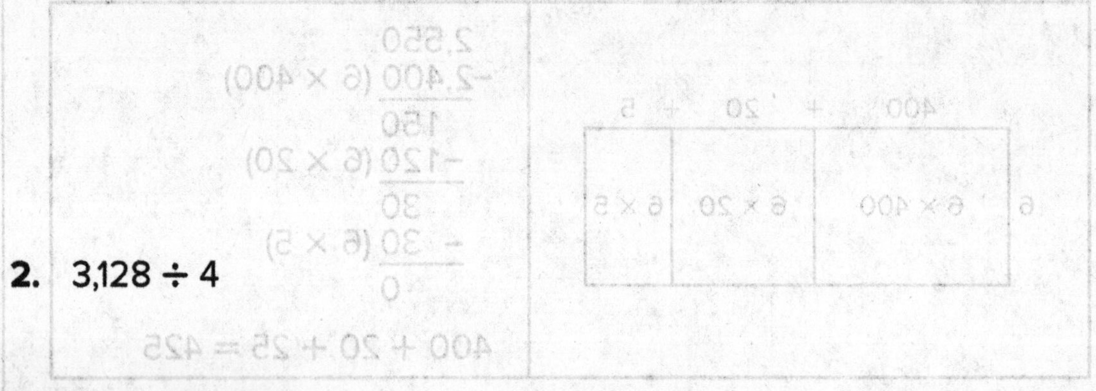

2. 3,128 ÷ 4

3. 4,068 ÷ 9

4. 1,974 ÷ 6

Understand Remainders

Name ______________________

Review

You can use partial quotients to divide and find a remainder.

324 ÷ 5 = ?

```
  324
- 300 (5× 60)
   24
 - 20 (5× 4)
    4
```

324 ÷ 5 = 64 R4

Use partial quotients to find the quotient and remainder.

1. 185 ÷ 7 = ?

2. 329 ÷ 4 = ?

Use partial quotients to solve.

3. Marcy has 187 stickers to share equally with 8 friends. How many stickers will she give to each friend? How many stickers will be left over?

4. Harold has 1,223 vocabulary words to learn. He plans to learn an equal number of words each month for 6 months. How many words will he learn each month? How many additional words will he need to learn?

Understand Remainders

Name ______________________________

Write a word problem that may be solved by calculating the quotient. Include a question that requires an understanding of the remainder. Then use partial quotients to solve.

1. 257 ÷ 8

2. 347 ÷ 3

3. 2,658 ÷ 9

4. 1,784 ÷ 6

Make Sense of a Remainder

Name ______________________________

Review

You can use a remainder to solve a real-world problem.

Ben puts 365 crackers into bags. Each bag holds 8 crackers. How many bags will Ben need to hold all of the crackers?

$365 \div 8 = ?$ $\begin{array}{r} 365 \\ -\ 320 \\ \hline 45 \\ -\ 40 \\ \hline 5 \end{array}$ $\begin{array}{l} \\ (8 \times 40) \\ \\ (8 \times 5) \\ \\ \end{array}$	$365 \div 8 = 45$ R5 Ben will need another bag for the leftover crackers. He will need 46 bags.

Write a division equation to represent the problem. Then solve.

1. Alex puts 27 candles into boxes. Each box holds 4 candles. How many boxes can Alex fill?

2. Sandra puts 167 light bulbs into boxes. Each box holds 6 light bulbs. How many boxes will Sandra need to hold all of the light bulbs?

3. Kim has 74 nickels to give to her 5 friends. She plans to give an equal number of nickels to each friend. If each friend receives the maximum number of nickels, how many nickels will be left over?

4. George equally divides 23 dog biscuits among 2 dogs. How many dog biscuits will each dog receive?

5. Jerry has a phone card with 214 minutes of talk time available. He plans to use the same number of minutes each day for 7 days. What is the maximum number of minutes he can use each day?

Make Sense of a Remainder

Name ______________________________

Fill in a 2- or 3-digit number and a 1-digit number to make a division problem that has a remainder. Then solve.

1. John has __________ marbles to put into bags. Each bag will hold __________ marbles. How many bags will he need to hold all of the marbles?

2. Matt puts __________ ornaments into boxes. Each box holds __________ ornaments. How many boxes can Matt fill?

3. A bakery is giving away __________ muffins to __________ local schools. Each school will receive an equal number of muffins. How many muffins will be left over?

4. A principal equally divides __________ inches of ribbon among __________ teachers. How many inches of ribbon will each teacher receive?

5. Francis has __________ bolts to give to __________ construction workers. She plans to give an equal number of bolts to each worker. If each worker receives the maximum number of bolts, how many bolts will be left over?

6. A total of __________ oranges will be equally divided among __________ groups of students. How many oranges will each group receive?

Solve Multi-Step Problems Using Division

Name

Review

You can solve multi-step problems involving division in steps.

Herman is packaging dog treats. There will be 8 dog treats in each bag. Herman had 219 dog treats, but he gave 14 dog treats to his best friend. How many bags of dog treats can Herman fill?

Step 1: Find out how many dog treats he has left after giving some to his friend.	Step 2: Divide to find out how many bags he can fill.
$219 - 14 = 205$	$205 \div 8 = b$ 205 − 160 (8 × 20) 45 − 40 (8 × 5) 5

$205 \div 8 = 25$ R5

Herman can fill 25 bags.

Solve. Show your work.

1. A mother buys 51 collectible action figures. She gives 5 action figures to her friend's daughter. With the remaining action figures, she plans to give an equal number to each of her 4 children. How many action figures will be left over?

2. Samuel has 124 post cards. He buys 9 more post cards. He plans to put the postcards into envelopes. Each envelope will hold 6 postcards. How many envelopes can Samuel fill?

Solve Multi-Step Problems Using Division

Name ______________________________

Fill in a 3-digit number and a 1-digit number to make a division problem that has a remainder. Solve the problem. After solving, write another question that is related to the problem and solve.

1. Thomas has ____________ tennis balls. He buys 4 more tennis balls. He places the tennis balls into boxes. Each box will hold ____________ tennis balls. How many boxes will Thomas fill?

2. A school gym has ____________ fitness bands. The gym teacher buys 7 more fitness bands. She plans to place an equal number of fitness bands in each of ____________ bags. How many fitness bands will be left over?

3. A school coach has ____________ basketballs. The coach gives 5 basketballs to a community center. The coach will store the remaining basketballs in metal crates. Each crate holds ____________ basketballs. How many crates will the coach need to hold all of the basketballs?

4. A physical education teacher has ____________ jump ropes for her students to use throughout the day. The teacher throws away 2 of the jump ropes. With the remaining jump ropes, the teacher plans to put an equal number of jump ropes into ____________ boxes. How many jump ropes will be placed in each box?

Equivalent Fractions

Name ______________________________

Review

You can use fraction strips to determine if two fractions are equivalent.

Are the fractions, $\frac{1}{5}$ and $\frac{2}{10}$, equivalent?

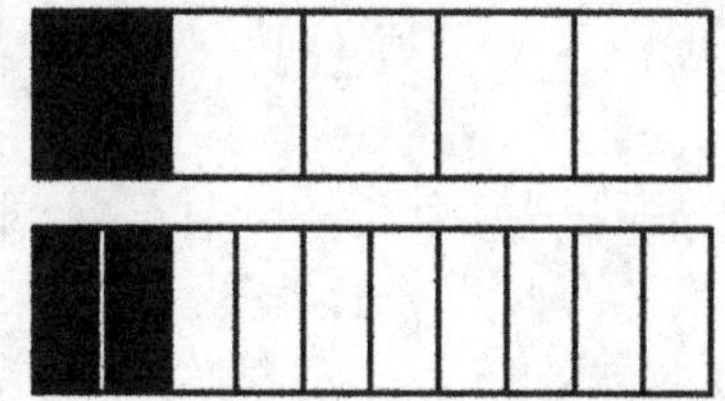

Yes, they are equivalent because they represent the same amount.

Decide if the fractions are equivalent. Write yes or no. Use a fraction model to justify your answer.

1. $\frac{1}{4}$ and $\frac{2}{8}$

2. $\frac{2}{3}$ and $\frac{6}{8}$

3. $\frac{1}{3}$ and $\frac{4}{12}$

Use a fraction model to solve.

4. On Monday, Angie decorated $\frac{1}{3}$ of a set of cards. On Tuesday, she decorated $\frac{2}{6}$ of the same set of cards. Did she decorate the same number of cards on both days?

5. Jason and Matt completed the same amount of a set of homework problems. Jason divided his assignment into 6 equal parts, and Matt divided his assignment into 8 equal parts. How many parts might they have each completed?

Lesson **8-1 • Extend Thinking**

Equivalent Fractions

Name ________________________________

Use a fraction model to determine if the fractions are equivalent. If the fractions are equivalent, write a real-world problem involving the fractions. If the fractions are not equivalent, explain why.

1. $\frac{4}{6}$ and $\frac{8}{12}$

2. $\frac{4}{5}$ and $\frac{2}{3}$

3. $\frac{3}{4}$ and $\frac{6}{8}$

4. $\frac{2}{5}$ and $\frac{6}{10}$

Generate Equivalent Fractions Using Models

Name ______________________________

Review

You can create a model to show the multiplication or division used to find an equivalent fraction.

$$\frac{6}{8} = \frac{?}{4}$$

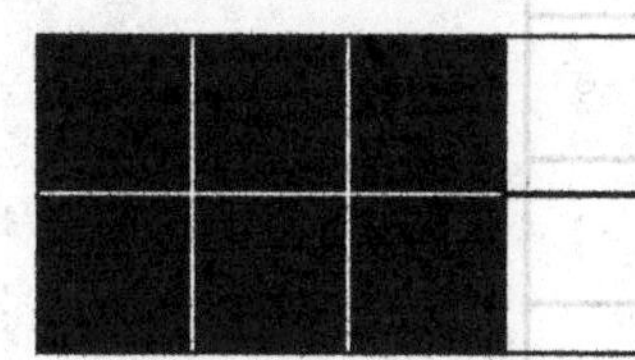

$$\frac{6}{8} = \frac{6 \div 2}{8 \div 2} = \frac{3}{4}$$

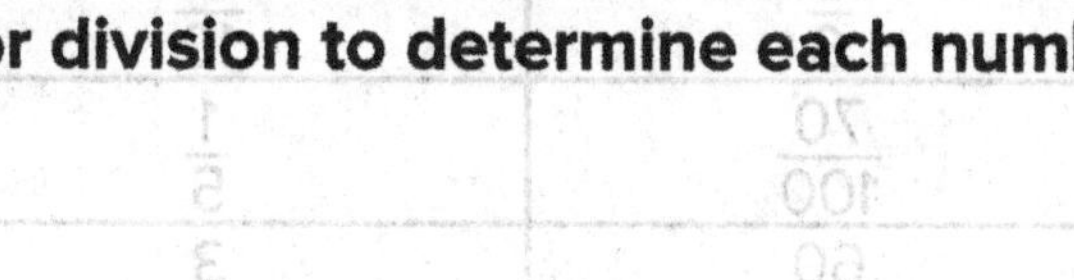

Find the missing number in each set of equivalent fractions. Explain how you used multiplication or division to determine each number.

1. $\frac{3}{5} = \frac{?}{10}$

2. $\frac{4}{10} = \frac{?}{100}$

3. $\frac{6}{12} = \frac{?}{4}$

Use a fraction model to solve.

4. Lucas and Kevin bake muffins using the same size muffin kit. Lucas eats $\frac{2}{6}$ of the muffins he bakes. Kevin eats the same amount, but his muffins are smaller. Kevin makes 12 muffins with his box. What fraction of the muffins did Kevin eat?

Generate Equivalent Fractions Using Models

Name ______________________________

Match each fraction on the left with the equivalent fraction on the right.

1.	$\frac{2}{10}$	$\frac{4}{10}$
2.	$\frac{4}{6}$	$\frac{50}{100}$
3.	$\frac{6}{8}$	$\frac{1}{6}$
4.	$\frac{1}{2}$	$\frac{6}{10}$
5.	$\frac{1}{4}$	$\frac{8}{12}$
6.	$\frac{2}{5}$	$\frac{8}{10}$
7.	$\frac{2}{12}$	$\frac{10}{12}$
8.	$\frac{5}{6}$	$\frac{7}{10}$
9.	$\frac{70}{100}$	$\frac{1}{5}$
10.	$\frac{60}{100}$	$\frac{3}{12}$
11.	$\frac{4}{5}$	$\frac{90}{100}$
12.	$\frac{9}{10}$	$\frac{3}{4}$

Find an equivalent fraction. Explain the process you used to find the equivalent fraction. Then, write a statement that involves a real-world context to show the equality of the fractions.

13. $\frac{5}{6}$

14. $\frac{80}{100}$

Generate Equivalent Fractions Using Number Lines

Name ______________________

Review

You can use a number line to find an equivalent fraction.

$$\frac{3}{5} = \frac{?}{10}$$

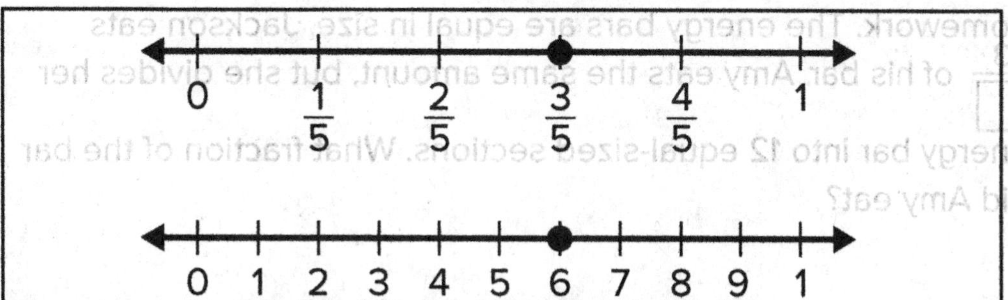

$$\frac{3}{5} = \frac{3 \times 2}{5 \times 2} = \frac{6}{10}$$

Find the missing number in each set of equivalent fractions. Explain how you used multiplication or division to determine each number.

1. $\frac{4}{6} = \frac{?}{3}$

2. $\frac{4}{5} = \frac{?}{10}$

3. $\frac{2}{8} = \frac{?}{4}$

Use a number line to solve.

4. Kim and Tonya are given the same size piece of fabric. Kim uses $\frac{1}{3}$ of the piece of fabric. Tonya uses the same amount, but her piece of fabric is divided into 6 smaller squares. What fraction of the fabric did Tonya use?

Generate Equivalent Fractions Using Number Lines

Name ______________________________

Fill in the blank to complete each problem to show equivalent fractions. Use a number line to represent the problem. Then solve.

1. Jackson and Amy are given energy bars to eat while doing their homework. The energy bars are equal in size. Jackson eats $\frac{3}{\square}$ of his bar. Amy eats the same amount, but she divides her energy bar into 12 equal-sized sections. What fraction of the bar did Amy eat?

2. Kaitlin and Joshua are both given the same homework assignment. Kaitlin has finished $\frac{4}{\square}$ of the assignment. Joshua has finished the same amount of the assignment, but he decides to divide the assignment into 10 equal-sized sections. What fraction of the assignment has Joshua finished?

3. Camille and Sarah take a break while doing homework. The breaks have an equal number of minutes. Camille spends $\frac{3}{\square}$ of her break meditating. Sarah spends the same amount of time meditating, but she divides her break into 4 equal chunks of time. What fraction of the break did Sarah spend meditating?

Compare Fractions Using Benchmarks

Name ______________________________

Review

You can use the benchmarks, 0, $\frac{1}{2}$, and 1 to compare fractions.

Compare $\frac{6}{10}$ and $\frac{5}{6}$.

Close to 1

Close to $\frac{1}{2}$

So, $\frac{6}{10} < \frac{5}{6}$.

Compare the fractions using benchmarks. Write >, <, or = to record the comparison. Explain your thinking.

1. $\frac{3}{5}$ ____ $\frac{7}{8}$

2. $\frac{2}{4}$ ____ $\frac{5}{10}$

3. $\frac{4}{6}$ ____ $\frac{2}{8}$

4. $\frac{3}{12}$ ____ $\frac{2}{8}$

5. $\frac{10}{12}$ ____ $\frac{6}{10}$

Compare Fractions Using Benchmarks

Name ____________________

Write a word problem that involves comparing the fractions. Then solve. Explain how you used benchmark numbers to arrive at your answer.

1. $\frac{3}{5}$ and $\frac{6}{8}$

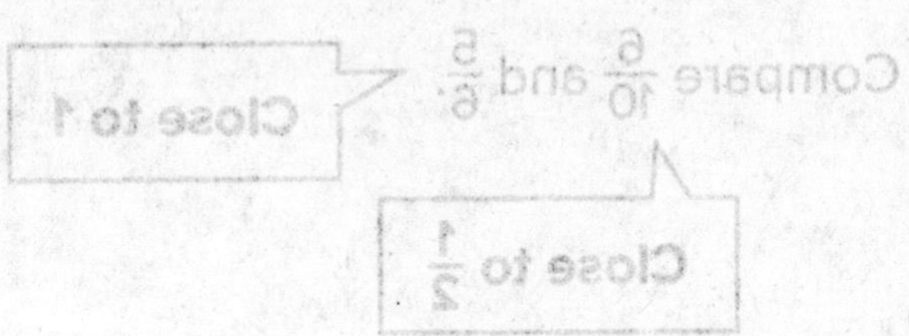

2. $\frac{4}{10}$ and $\frac{5}{6}$

3. $\frac{11}{12}$ and $\frac{5}{8}$

4. $\frac{2}{6}$ and $\frac{4}{10}$

Other Ways to Compare Fractions

Name ______________________________

Review

You can use equivalent fractions to compare fractions with unlike denominators.

Compare $\frac{2}{3}$ and $\frac{3}{6}$. $\downarrow \quad \downarrow$ $\frac{4}{6} > \frac{3}{6}$	You can create equivalent fractions with like denominators.

Four $\frac{1}{6}$ parts > three $\frac{1}{6}$ parts.

Use equivalent fractions with like denominators to solve.
Write >, <, or = to record the comparison. Explain your thinking.

1. $\frac{3}{5}$ ____ $\frac{5}{10}$

2. $\frac{3}{4}$ ____ $\frac{6}{8}$

Use equivalent fractions with like numerators to solve.
Write >, <, or = to record the comparison. Explain your thinking.

3. $\frac{2}{3}$ ____ $\frac{4}{10}$

4. $\frac{8}{12}$ ____ $\frac{4}{5}$

Other Ways to Compare Fractions

Name ______________________________

Write a word problem that involves comparing the fractions. Then solve. Explain how you used equivalent fractions to arrive at your answer.

1. $\frac{3}{8}$ and $\frac{2}{4}$

2. $\frac{6}{10}$ and $\frac{3}{5}$

3. $\frac{9}{12}$ and $\frac{5}{6}$

4. $\frac{2}{5}$ and $\frac{4}{12}$

Understand Decomposing Fractions

Name ______________________________

Review

You can decompose a fraction into smaller fractions.

You can use unit fractions to decompose a fraction into the greatest number of parts.	You can combine unit fractions to show more than one way to make greater addends.
$\frac{5}{4}=\frac{1}{4}+\frac{1}{4}+\frac{1}{4}+\frac{1}{4}+\frac{1}{4}$	$\frac{5}{4}=\frac{2}{4}+\frac{2}{4}+\frac{1}{4}$ $\frac{5}{4}=\frac{3}{4}+\frac{2}{4}$ $\frac{5}{4}=\frac{4}{4}+\frac{1}{4}$

What unit fractions decompose the fraction?

1. $\frac{2}{3}=$ ______ + ______

2. $\frac{5}{6}=$ ______ + ______ + ______ + ______ + ______

What fractions decompose the fraction into 3 parts?

3. $\frac{11}{8}=$ ______ + ______ + ______

4. $\frac{17}{12}=$ ______ + ______ + ______

What missing fractions decompose the fraction into 5 parts?

5. $\frac{7}{6}=\frac{2}{6}+\frac{2}{6}+\frac{1}{6}+$ ______ + ______

6. $\frac{13}{10}=\frac{5}{10}+\frac{5}{10}+\frac{1}{10}+$ ______ + ______

Understand Decomposing Fractions

Name ______________________

Write a word problem that involves decomposition of the fraction. Then solve. Use equations and drawings to show your work.

1. $\frac{4}{5}$

2. $\frac{9}{10}$

3. $\frac{7}{8}$

Represent Adding Fractions

Name ______________________________

Review

You can use representations to add fractions.

You can use fraction strips to add.

$\frac{2}{10} + \frac{1}{10} + \frac{3}{10} + \frac{4}{10} = ?$

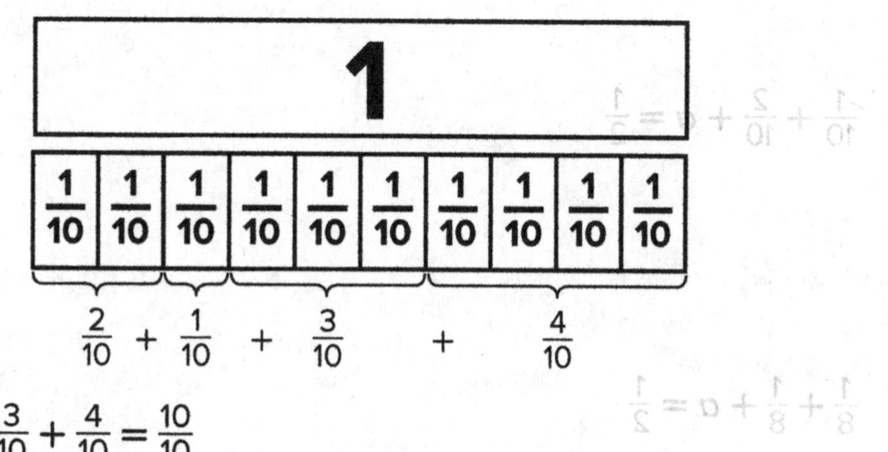

$\frac{2}{10} + \frac{1}{10} + \frac{3}{10} + \frac{4}{10} = \frac{10}{10}$

What is the sum? Complete the fraction model to represent the equation.

1. $\frac{2}{5} + \frac{1}{5} + \frac{1}{5} =$ ______

0 $\frac{1}{5}$ $\frac{2}{5}$ $\frac{3}{5}$ $\frac{4}{5}$ $\frac{5}{5}$

2. $\frac{3}{8} + \frac{2}{8} + \frac{3}{8} =$ ______

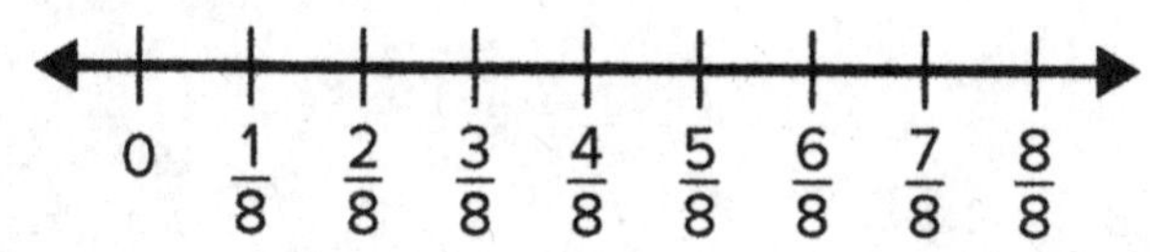

Lesson **9-2 • Extend Thinking**

Represent Adding Fractions

Name ______________________________

What missing fraction makes the equation true? Use a number line to show your work.

1. $\frac{2}{12} + \frac{3}{12} + a = \frac{3}{4}$

2. $\frac{1}{10} + \frac{2}{10} + a = \frac{1}{2}$

3. $\frac{1}{8} + \frac{1}{8} + a = \frac{1}{2}$

4. $\frac{1}{6} + \frac{1}{6} + a = \frac{2}{3}$

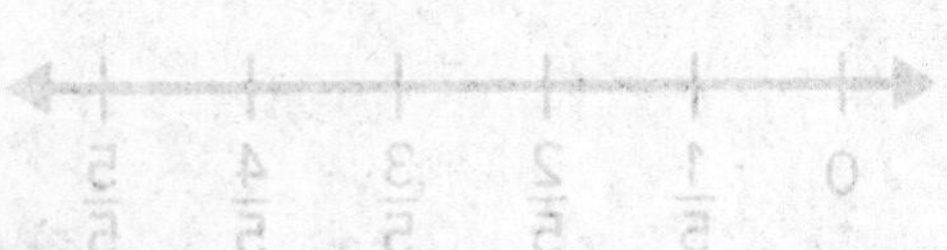

5. $\frac{1}{12} + \frac{4}{12} + a = \frac{2}{3}$

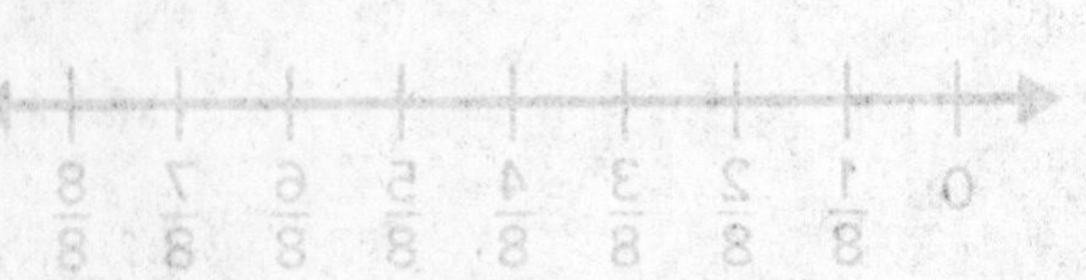

Add Fractions with Like Denominators

Name ______________________________

Review

To add fractions with the same denominator, you add the numerators and the denominator stays the same.

$\frac{7}{5} + \frac{2}{5} = ?$

You can use a number line to show this sum.

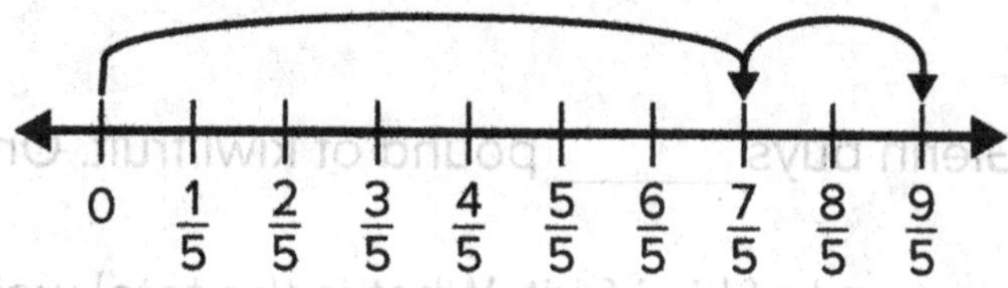

$\frac{7}{5} + \frac{2}{5} = \frac{9}{5}$

What is the missing value that completes the equation?

1. $\frac{6}{8} + \frac{2}{8} = \frac{}{8}$

2. $\frac{4}{5} + \frac{6}{5} = \frac{}{5}$

3. $\frac{7}{10} + \frac{4}{10} = \frac{}{10}$

4. ______ $+ \frac{1}{4} = \frac{3}{4}$

5. $\frac{2}{12} +$ ______ $= \frac{9}{12}$

6. Mike has $\frac{3}{4}$ cup of corn meal. His friend, Eric, gives him an additional $\frac{2}{4}$ cup of corn meal. How much corn meal does Mike have now?

Add Fractions with Like Denominators

Name ____________________

Fill in two fractions with the same denominator to make an addition problem. Solve. Use a representation to show the sum.

1. Kylie has ______ cup of flour. Her mom gives her ______ cup more flour. How much flour does Kylie have now?

2. On Friday, Glenn buys ______ pound of kiwi fruit. On Monday, he buys ______ pound of kiwi fruit. What is the total weight of the kiwi fruit bought by Glenn on Friday and Monday?

3. To prepare bakery items for the week, Elizabeth used ______ gallon of skim milk and ______ gallon of whole milk. What is the total amount of milk that Elizabeth used this week?

4. Over the weekend, Harold drank ______ liter of orange juice and ______ liter of tea. What is the total amount of orange juice and tea that Harold drank over the weekend?

Represent Subtracting Fractions

Name ______________________________

Review

You can use representations to subtract fractions.

You can use fraction strips to subtract.

$\frac{5}{6} - \frac{2}{6} = ?$

First, show $\frac{5}{6}$ of a whole. Next, cross out 2 of the $\frac{1}{6}$ strips.

$\frac{1}{6}$	$\frac{1}{6}$	$\frac{1}{6}$	~~$\frac{1}{6}$~~	~~$\frac{1}{6}$~~	

$\frac{5}{6} - \frac{2}{6} = \frac{3}{6}$

What is the difference? Complete the fraction model to represent the equation.

1. $\frac{4}{5} - \frac{1}{5} =$ ______

$\frac{1}{5}$	$\frac{1}{5}$	$\frac{1}{5}$	$\frac{1}{5}$	

2. $\frac{6}{10} - \frac{4}{10} =$ ______

$\frac{1}{10}$	$\frac{1}{10}$	$\frac{1}{10}$	$\frac{1}{10}$	$\frac{1}{10}$	$\frac{1}{10}$				

3. $\frac{11}{12} - \frac{5}{12} =$ ______

$\frac{1}{12}$	$\frac{1}{12}$	$\frac{1}{12}$	$\frac{1}{12}$	$\frac{1}{12}$	$\frac{1}{12}$	$\frac{1}{12}$	$\frac{1}{12}$	$\frac{1}{12}$	$\frac{1}{12}$	$\frac{1}{12}$	

Solve. Use an equation and fraction model to show your work.

4. Donna has $\frac{2}{8}$ of a bottle of glitter left over. She started with a full bottle of glitter. How much glitter did Donna use?

Lesson **9-4 • Extend Thinking**

Represent Subtracting Fractions

Name ____

What missing fraction makes the equation true? Use a number line to show your work.

1. $\frac{8}{12} - a = \frac{1}{4}$

2. $\frac{9}{10} - a = \frac{1}{2}$

3. $\frac{5}{6} - a = \frac{1}{3}$

4. $\frac{7}{8} - a = \frac{3}{4}$

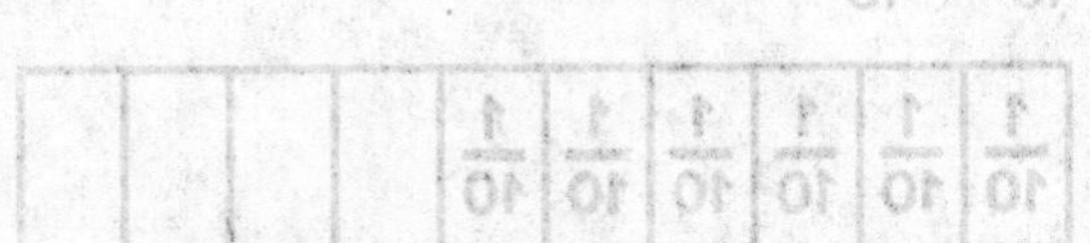

5. $\frac{6}{12} - a = \frac{1}{6}$

Subtract Fractions with Like Denominators

Name ______________________________

Review

To subtract fractions with the same denominator, you subtract the numerators and the denominator stays the same.

$\frac{4}{5} - \frac{2}{5} = ?$

You can use a number line to show this difference.

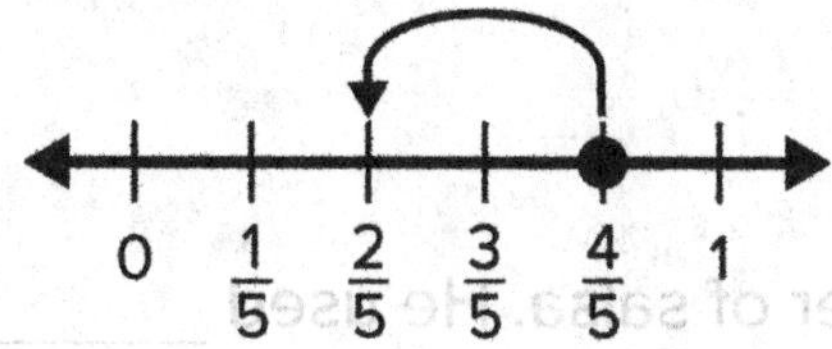

$\frac{4}{5} - \frac{2}{5} = \frac{2}{5}$

What is the missing value that completes the equation?

1. $\frac{5}{6} - \frac{2}{6} = \frac{\quad}{6}$

2. $\frac{9}{12} - \frac{6}{12} = \frac{\quad}{12}$

3. $\frac{8}{10} - \frac{2}{10} = \frac{\quad}{10}$

4. ______ $- \frac{3}{8} = \frac{2}{8}$

5. $\frac{17}{100} -$ ______ $= \frac{13}{100}$

Solve. Use an equation and representation to show your work.

6. On Wednesday, Jill had $\frac{10}{12}$ of a homework assignment left to complete. The next day, Jill completed $\frac{4}{12}$ of the assignment. How much of the assignment does Jill still need to complete?

Subtract Fractions with Like Denominators

Name ____

Fill in two fractions with the same denominator to make a subtraction problem. Solve. Use a representation to show the difference.

1. Ashton had ______ pound of cheese. He used ______ pound of cheese to make enchiladas. How much cheese does Ashton have left?

2. Ben had ______ liter of salsa. He used ______ liter of salsa for a birthday party. How much salsa does Ben have now?

3. Amy had ______ of a lasagna. Her friends ate ______ of the lasagna. How much of the lasagna does Amy have left?

4. Isabella had ______ pound of guacamole at the start of the week.

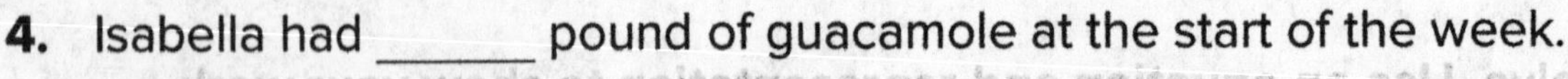

By the end of the week, she had ______ pound left. How much guacamole did she eat during the week?

Solve Problems Involving Fractions

Name ________________________________

Review

You can solve a word problem involving fractions by writing out steps.

On a vacation, Tim took $\frac{3}{4}$ pound of dried cranberries. He took $\frac{2}{4}$ pound less dried bananas than dried cranberries. What is the total weight of the dried fruit that Tim took on his vacation?

Step 1: Subtract to find the weight of the dried bananas.

$\frac{3}{4} - \frac{2}{4} = \frac{1}{4}$

Step 2: Add to find the total weight of the fruit.

$\frac{3}{4} + \frac{1}{4} = \frac{4}{4}$ The total weight of the dried fruit is $\frac{4}{4}$ pound or 1 pound.

Solve. Use equations and a representation to show your work.

1. This morning, Thomas ran $\frac{7}{10}$ mile. This afternoon, he ran $\frac{3}{10}$ mile less than the distance he ran this morning.

Thomas ran ________ mile this afternoon. What is the total distance that Thomas ran today?

2. Joyce has used $\frac{8}{10}$ liter of peppermint shampoo. She has used $\frac{2}{10}$ liter less coconut shampoo than peppermint shampoo.

Joyce used ________ liter of coconut shampoo. How much shampoo has Joyce used?

Solve Problems Involving Fractions

Name ______________________________

Solve each problem. Use equations and two different representations to show your work.

1. Estelle ate $\frac{3}{8}$ pound of swiss cheese. She ate $\frac{2}{8}$ pound more provolone cheese than swiss cheese. How much cheese did Estelle eat?

2. Melody has used $\frac{7}{10}$ liter of ketchup. She has used $\frac{3}{10}$ liter less mustard than ketchup. How much ketchup and mustard has she used?

3. Robin ate $\frac{3}{12}$ of a pizza on Monday. She ate $\frac{2}{12}$ more pizza on Tuesday than Monday. How much pizza did Robin eat on Monday and Tuesday?

4. Carmen drank $\frac{8}{10}$ liter of grape flavored water. She drank $\frac{3}{10}$ liter less watermelon flavored water than grape flavored water. How much flavored water did Carmen drink?

5. Sharon ate $\frac{5}{8}$ of a loaf of wheat bread. She ate $\frac{3}{8}$ of a loaf less white bread than wheat bread. If the loaves are the same size, how much of a loaf of bread did Sharon eat?

Understand Decomposing Mixed Numbers

Name ______________________________

Review

You can decompose a mixed number into whole number parts and fraction parts.

$2\frac{5}{6}$

$2\frac{5}{6} = 1 + 1 + \frac{5}{6}$

$2\frac{5}{6} = \frac{6}{6} + \frac{6}{6} + \frac{5}{6}$

How can you decompose the mixed number? Match equations to represent the decompositions.

1. $1\frac{4}{5}$

2. $2\frac{1}{5}$

3. $2\frac{3}{5}$

4. $1\frac{2}{5}$

Understand Decomposing Mixed Numbers

Name ______________________________

Write a word problem that involves decomposition of the mixed number. Then solve. Use an equation and drawing to show your work.

1. $2\frac{2}{3}$

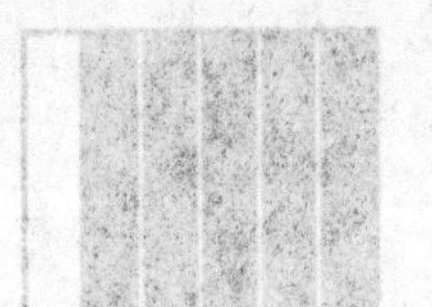

2. $2\frac{3}{8}$

3. $4\frac{7}{12}$

Represent Adding Mixed Numbers

Name ________________________________

Review

You can use representations to add mixed numbers.

You can use fraction strips to add.

$1\frac{2}{5} + 3\frac{4}{5} = ?$

Arrange the fraction strips to group the whole number parts and the fraction parts. Then, use 5 of the $\frac{1}{5}$ fraction parts to make another whole.

1	1	1	1	$\frac{1}{5}$	$\frac{1}{5}$	$\frac{1}{5}$	$\frac{1}{5}$	$\frac{1}{5}$	$\frac{1}{5}$

1	1	1	1	1	$\frac{1}{5}$

$1\frac{2}{5} + 3\frac{4}{5} = 5\frac{1}{5}$

What is the sum? Use a representation to show your work.

1. $1\frac{2}{4} + 2\frac{1}{4} = \square$

2. $2\frac{3}{6} + 1\frac{3}{6} = \square$

3. $3\frac{3}{8} + 2\frac{7}{8} = \square$

4. $2\frac{8}{12} + 3\frac{9}{12} = \square$

Represent Adding Mixed Numbers

Name ______________________________

What missing mixed number makes the equation true? Use a number line to show your work.

1. $1\frac{2}{12} + 2\frac{3}{12} + a = 4\frac{2}{3}$

2. $1\frac{1}{10} + 2\frac{2}{10} + a = 4\frac{1}{2}$

3. $2\frac{2}{8} + 1\frac{3}{8} + a = 5\frac{3}{4}$

4. $3\frac{2}{12} + 2\frac{5}{12} + a = 8$

Add Mixed Numbers

Name ______________________________

Review

You can use equivalent fractions to add mixed numbers.

$1\frac{3}{4} + 1\frac{2}{4} = ?$

$1\frac{3}{4} + 1\frac{2}{4} = \frac{7}{4} + \frac{6}{4}$ — The sum is a fraction that is greater than 1.

$= \frac{13}{4}$

$= 3\frac{1}{4}$

What are the missing values that complete the equation?

1. $2\frac{3}{8} + 1\frac{4}{8} = 2 + \frac{3}{8} + \square + \frac{\square}{\square}$

$= 2 + 1 + \frac{3}{8} + \frac{\square}{\square}$

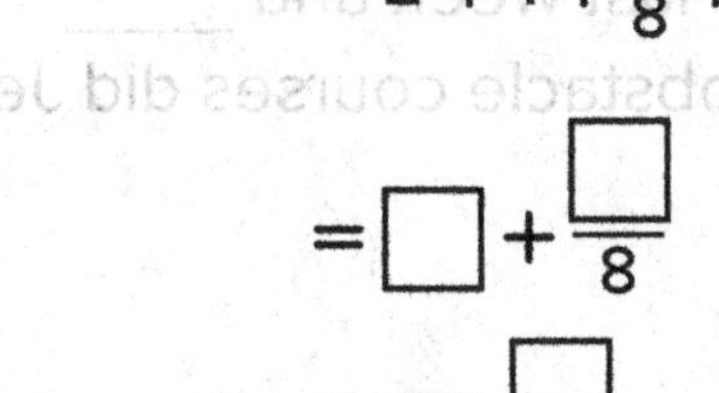

$= \square + \frac{\square}{8}$

$= \square\frac{\square}{\square}$

2. $1\frac{4}{10} + 2\frac{7}{10} = \frac{\square}{10} + \frac{\square}{10}$

$= \frac{\square}{10}$

$= \square + \frac{\square}{10}$

$= \square\frac{\square}{\square}$

3. $3\frac{6}{12} + 2\frac{11}{12} = \frac{\square}{12} + \frac{\square}{12}$

$= \frac{\square}{12}$

$= \square + \frac{\square}{12}$

$= \square\frac{\square}{\square}$

4. $2\frac{3}{6} + 3\frac{5}{6} = \frac{\square}{6} + \frac{\square}{6}$

$= \frac{\square}{6}$

$= \square + \frac{\square}{6}$

$= \square\frac{\square}{\square}$

Add Mixed Numbers

Name ______________________________

Fill in two mixed numbers with the same denominator to make an addition problem. Solve. Use equations to show your work.

1. Sheila swam _____ laps on Tuesday. She swam _____ laps on Wednesday. How many laps did Sheila swim on Tuesday and Wednesday?

2. Deborah hiked _____ miles on Friday. She hiked _____ miles on Saturday. What is the total distance that Deborah hiked on Friday and Saturday?

3. Jerry completed _____ obstacle courses last week and _____ obstacle courses this week. How many obstacle courses did Jerry completed over the past two weeks?

4. Jim spent _____ minutes climbing on the climbing wall and _____ minutes playing basketball. How many minutes did Jim spend climbing on the climbing wall and playing basketball?

Represent Subtracting With Mixed Numbers

Name ________________________________

Review

You can use representations to subtract with mixed numbers.

You can use fraction strips to subtract.

$4\frac{2}{5} - 1\frac{3}{5} = ?$

Subtract the whole number part of $1\frac{3}{5}$ from $4\frac{2}{5}$.

1	1	1	~~1~~	$\frac{1}{5}$	$\frac{1}{5}$

Decompose a whole to make more $\frac{1}{5}$ parts.

1	1	$\frac{1}{5}$	$\frac{1}{5}$	$\frac{1}{5}$	$\frac{1}{5}$	$\frac{1}{5}$	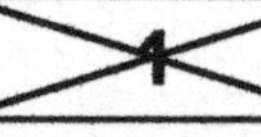	$\frac{1}{5}$	$\frac{1}{5}$

Now, subtract the fractional part of $1\frac{3}{5}$ from $4\frac{2}{5}$.

1	1	$\frac{1}{5}$	$\frac{1}{5}$	$\frac{1}{5}$	$\frac{1}{5}$	~~$\frac{1}{5}$~~	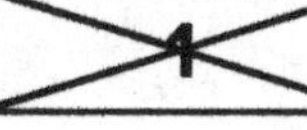	~~$\frac{1}{5}$~~	~~$\frac{1}{5}$~~

$4\frac{2}{5} - 1\frac{3}{5} = 2\frac{4}{5}$

What is the difference? Use a representation to show your work.

1. $4\frac{3}{4} - 2\frac{1}{4} = \square$

2. $3\frac{7}{8} - 1\frac{2}{8} = \square$

3. $2\frac{3}{6} - 1\frac{5}{6} = \square$

4. $4\frac{1}{5} - 2\frac{2}{5} = \square$

Represent Subtracting With Mixed Numbers

Name ______________________________

What missing mixed number makes the equation true? Use a number line to show your work.

1. $3\frac{4}{6} - a = 1\frac{1}{3}$

2. $4\frac{2}{8} - a = 2\frac{1}{2}$

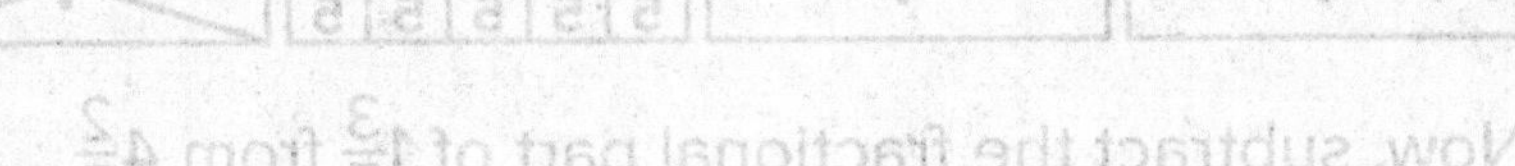

3. $2\frac{4}{10} - a = \frac{4}{5}$

4. $4\frac{6}{12} - a = 2\frac{3}{4}$

Subtract Mixed Numbers

Name ______________________________

Review

You can use a related addition equation to subtract mixed numbers.

$2\frac{1}{6} - 1\frac{3}{6} = p$

$1\frac{3}{6} + p = 2\frac{1}{6}$

Then, use the counting on method to count from $1\frac{3}{6}$ to $2\frac{1}{6}$.

$1\frac{3}{6} + \frac{3}{6} = 2$ — *You count on $\frac{4}{6}$ to get to $2\frac{1}{6}$.*

$2 + \frac{1}{6} = 2\frac{1}{6}$

$p = \frac{4}{6}$

What are the missing values that complete the equation?

1. $3\frac{4}{10} - 2\frac{1}{10} = p$

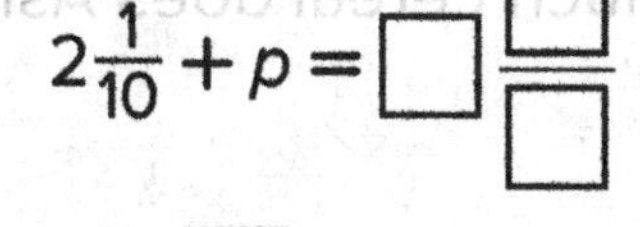

$2\frac{1}{10} + p = \square\frac{\square}{\square}$

$2\frac{1}{10} + \frac{\square}{10} = 3$

$3 + \frac{\square}{10} = 3\frac{4}{10}$

$p = \square\frac{\square}{\square}$

2. $4\frac{5}{12} - 2\frac{8}{12} = p$

$2\frac{8}{12} + p = \square\frac{\square}{\square}$

$2\frac{8}{12} + \frac{\square}{\square} = 4$

$4 + \frac{\square}{12} = 4\frac{5}{12}$

$p = \frac{\square}{\square}$

$p = \square\frac{\square}{\square}$

Subtract Mixed Numbers

Name ______________________________

Fill in two mixed numbers with the same denominator to make a subtraction problem. Solve. Use equations to show your work.

1. Kaylee had ______ cups of rice. She used ______ cups of rice. How much rice does Kaylee have now?

2. Yvonne made ______ liters of orange juice. She gave ______ liters of orange juice to a friend. How many liters of orange juice does Yvonne have left?

3. Ashley had ______ cups of cereal to eat on a vacation. She ate ____ cups of cereal during the vacation. How much cereal does Ashley have left?

4. Ben started the week with ______ gallons of milk. During the week, he drank ______ gallons of milk. How much milk does Ben have left?

Solve Problems Involving Mixed Numbers

Name ________________________________

Review

You can solve a word problem involving mixed numbers by writing out steps.

Carl bought 5 liters of water. He drank $1\frac{8}{10}$ liters of water on Monday and $2\frac{3}{10}$ liters on Tuesday. How much water was left?

Step 1: Add to find the amount of water that Carl drank on Monday and Tuesday.

$1\frac{8}{10} + 2\frac{3}{10} = 4\frac{1}{10}$

Step 2: Subtract to find the amount of water that was left.

$5 - 4\frac{1}{10} = \frac{9}{10}$ He had $\frac{9}{10}$ liter of water left.

Solve. Use equations and a drawing to show your work.

1. On Friday, Shelly used $3\frac{4}{8}$ cups of flour. On Saturday, she used $1\frac{6}{8}$ cups less flour than she used on Friday. What is the total amount of flour that Shelley used on Friday and Saturday?

   ______ + ______ = ______

2. Lindsey bought $2\frac{4}{6}$ pounds of bananas. She bought $1\frac{5}{6}$ pounds fewer oranges than bananas. How much fruit did Lindsey buy?

   ______ + ______ = ______

Solve Problems Involving Mixed Numbers

Name ______________________________

Solve each problem. Use equations and two different representations to show your work.

1. Victor ran $4\frac{3}{10}$ miles on Monday. He ran $1\frac{5}{10}$ miles more on Tuesday than he ran on Monday. How far did Victor run on Monday and Tuesday?

2. Joshua hiked $3\,\frac{7}{10}$ kilometers on Friday. He hiked $1\,\frac{8}{10}$ kilometers less on Saturday than Friday. How far did Joshua hike on Friday and Saturday?

3. On Wednesday, Sara jumped on the trampoline for $5\frac{3}{4}$ minutes. She rode her bike for $4\frac{1}{4}$ minutes more than the number of minutes spent jumping on the trampoline. How long did Sara jump on the trampoline and ride her bike?

4. Margo spent $3\frac{3}{6}$ minutes doing pushups and $4\frac{5}{6}$ minutes doing situps. She plans to spend a total of 15 minutes exercising. How many minutes does she have left to spend on other exercises?

Represent Multiplication of a Unit Fraction by a Whole Number

Name ______________________________

Review

You can use a fraction model and equations to represent multiplication of a unit fraction by a whole number.

$3 \times \frac{1}{4} = ?$

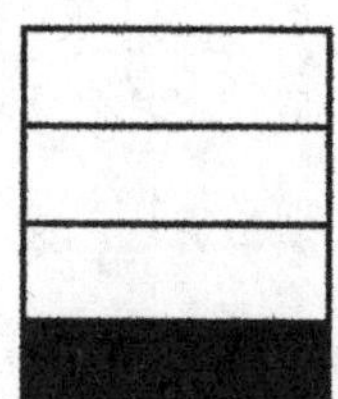
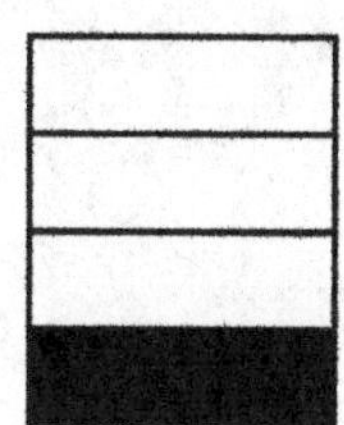

$\frac{1}{4} + \frac{1}{4} + \frac{1}{4} = \frac{3}{4}$

$3 \times \frac{1}{4} = \frac{3}{4}$

What missing value(s) completes the equation?

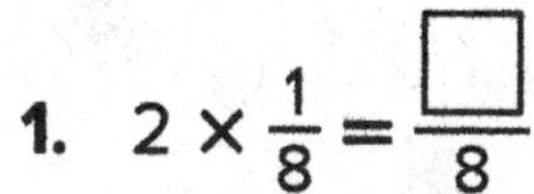

1. $2 \times \frac{1}{8} = \frac{\square}{8}$

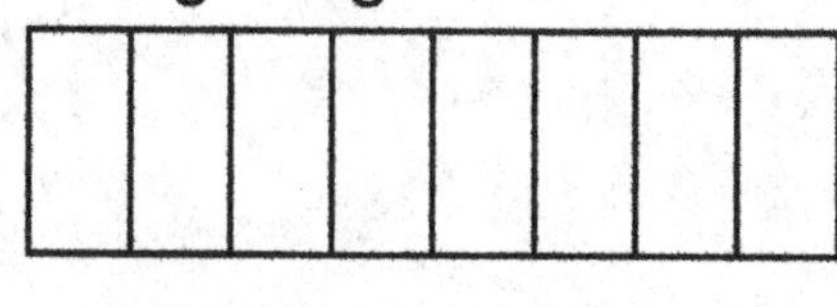
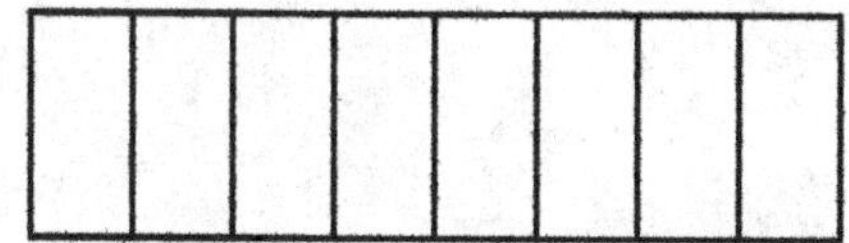

2. $9 \times \frac{1}{2} = \frac{\square}{2}$

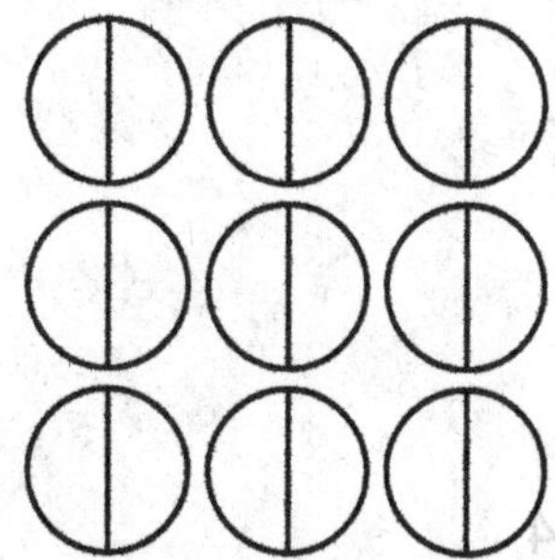

What missing value completes the equation.

3. $5 \times \frac{1}{4} = \frac{\square}{4}$

4. $7 \times \frac{1}{12} = \frac{\square}{12}$

5. $8 \times \frac{1}{2} = \frac{\square}{\square}$

6. $3 \times \frac{1}{10} = \frac{\square}{\square}$

Represent Multiplication of a Unit Fraction by a Whole Number

Name ______________________________

What missing unit fraction makes the equation true? Use a number line to show your work.

1. $8 \times a = 2$

2. $6 \times a = \frac{3}{4}$

3. $8 \times a = \frac{2}{3}$

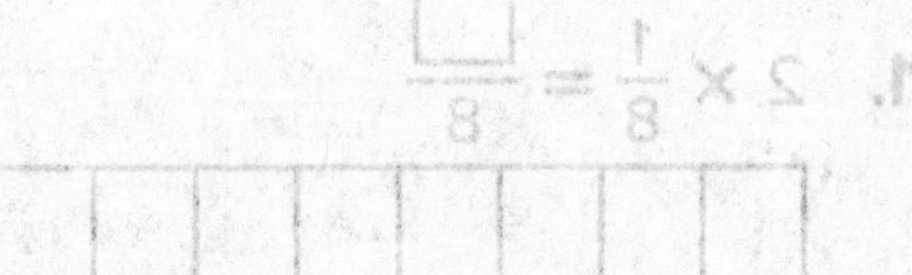

4. $4 \times a = \frac{2}{5}$

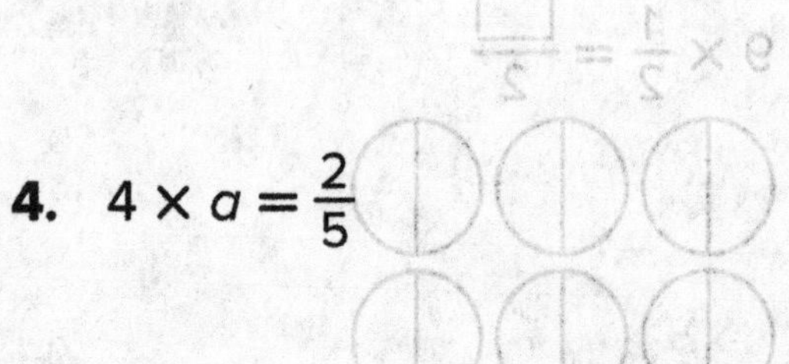

5. $16 \times a = \frac{4}{3}$

Understand Multiplying a Fraction by a Whole Number

Name ______________________________

Review

You can use a fraction model to understand multiplying a fraction by a whole number.

$4 \times \frac{3}{5} = ?$

$3 \times \frac{1}{5}$ $3 \times \frac{1}{5}$ $3 \times \frac{1}{5}$ $3 \times \frac{1}{5}$

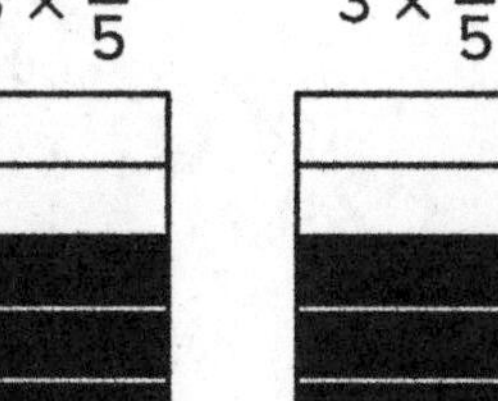
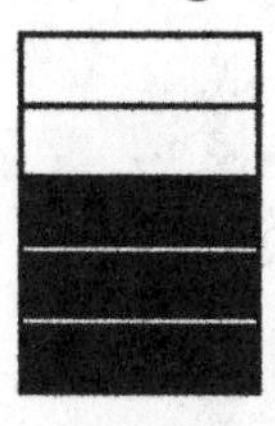
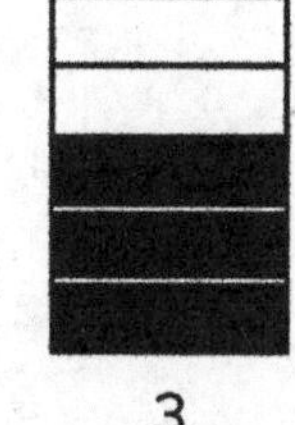

$\frac{3}{5}$ $\frac{3}{5}$ $\frac{3}{5}$ $\frac{3}{5}$

$$4 \times \frac{3}{5} = 4 \times (3 \times \frac{1}{5})$$
$$= 12 \times \frac{1}{5}$$
$$= \frac{12}{5}$$

What missing value completes the equation?

1. $2 \times \frac{2}{8} = \frac{\square}{8}$

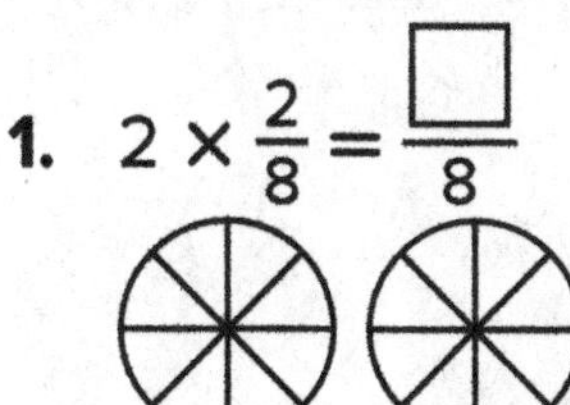

2. $3 \times \frac{3}{6} = \frac{\square}{6}$

3. $4 \times \frac{2}{5} = \frac{\square}{5}$

4. $2 \times \frac{5}{6} = \frac{\square}{6}$

Understand Multiplying a Fraction by a Whole Number

Name ____________________

What missing fraction makes the equation true? Use a number line to show your work.

1. $4 \times a = 3$

2. $6 \times a = \frac{3}{2}$

3. $10 \times a = 4$

4. $6 \times a = \frac{9}{5}$

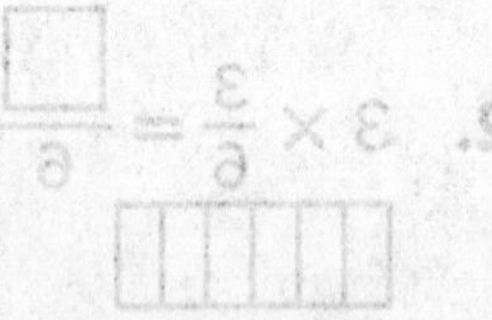

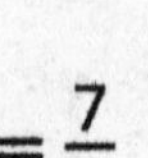

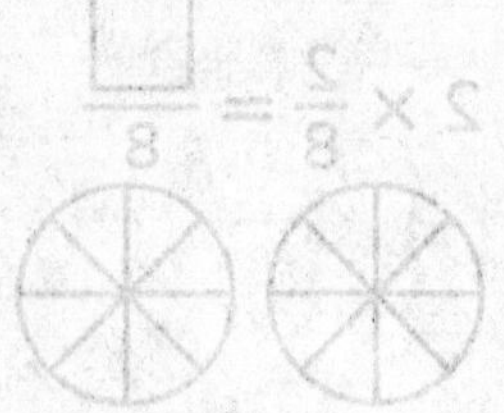

5. $2 \times a = \frac{7}{4}$

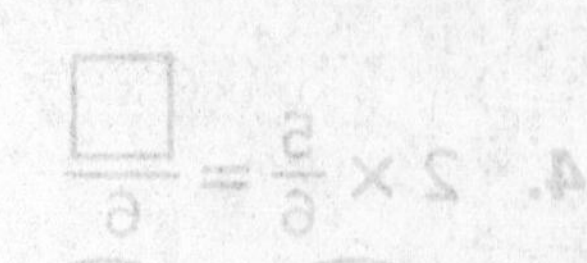

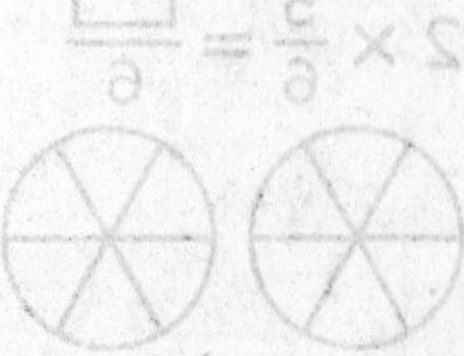

Lesson **11-3 • Reinforce Understanding**

Multiply a Fraction by a Whole Number

Name ______________________________

Review

You can use jumps on a number line to multiply.

$2 \times \frac{5}{6} = ?$

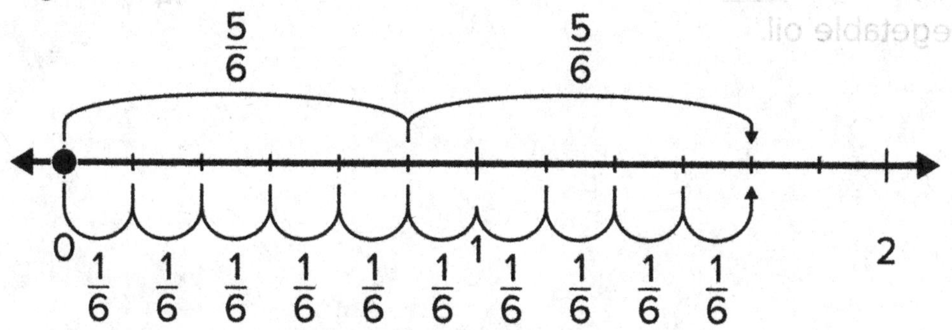

$2 \times (5 \times \frac{1}{6}) = ?$

$(2 \times 5) \times \frac{1}{6} = ?$

$10 \times \frac{1}{6} = \frac{10}{6}$

Use the Associative Property to group the factors, 2 and 5, instead.

What are the missing values that complete the equation?

1. $4 \times \frac{3}{4} = 4 \times$ ___ $\times \frac{1}{4}$

$=$ ___ $\times \frac{1}{4}$

$= \frac{\square}{4}$

$=$ ___

2. $2 \times \frac{7}{8} = 2 \times$ ___ $\times \frac{1}{8}$

$=$ ___ $\times \frac{1}{8}$

$= \frac{\square}{8}$

$=$ ___ $\frac{\square}{\square}$

Multiply a Fraction by a Whole Number

Name ______________________________

Which whole number and fraction make the statement true? Use equations to show your work.

1. Anna will use ____ cup of vegetable oil in each batch of pancakes. She makes ____ batches of pancakes. Anna uses $1\frac{2}{4}$ cups of vegetable oil.

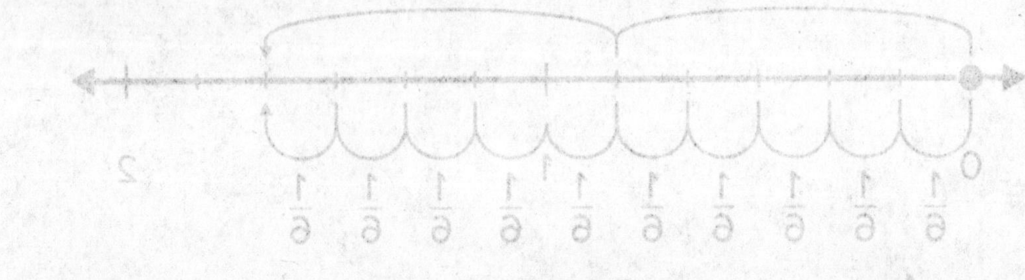

2. Randy will use ____ ounces of water in each casserole dish. He plans to make ____ casserole dishes. Randy uses $2\frac{4}{8}$ ounces of water.

3. John will use ____ cup of milk with each cake he bakes. He plans to bake ____ cakes. John uses 2 cups of milk.

Multiply a Mixed Number by a Whole Number

Name ______________________________

Review

You can use decomposition and the Distributive Property to multiply a mixed number by a whole number.

$3 \times 2\frac{5}{6} = ?$

$3 \times 2\frac{5}{6} = 3 \times (2 + \frac{5}{6})$

$= (3 \times 2) + (3 \times \frac{5}{6})$

$= 6 + \frac{15}{6}$

$= 6\frac{15}{6}$

$= 8\frac{3}{6}$

What are the missing values that complete the equation?

1. $4 \times 2\frac{3}{10} = 4 \times (2 + \frac{\square}{\square})$

$= (4 \times 2) + (4 \times \frac{\square}{\square})$

$= 8 + \frac{\square}{\square}$

$= 8\frac{\square}{\square}$

$= 9\frac{\square}{\square}$

2. $2 \times 3\frac{5}{6} = 2 \times (3 + \frac{\square}{\square})$

$= (2 \times 3) + (2 \times \frac{\square}{\square})$

$= 6 + \frac{\square}{\square}$

$= 6\frac{\square}{\square}$

$= 7\frac{\square}{\square}$

Multiply a Mixed Number by a Whole Number

Name ______________________________

Which mixed number and whole number make the statement true? Use equations to show your work.

1. Kenneth uses _____ milliliters of paint on each envelope that he decorates. He decorates _____ envelopes. Kenneth uses $13\frac{3}{5}$ milliliters of paint.

2. Tony uses _____ feet of yarn to make each hat. He plans to make _____ hats. Tony uses $12\frac{9}{12}$ feet of yarn to make the hats.

3. Joyce will use _____ pounds of decorative rocks to fill each vase. She will fill _____ vases. Joyce uses $26\frac{2}{8}$ pounds of rocks.

Solve Problems Involving Fractions and Mixed Numbers

Name ____________________

Review

You can solve a word problem involving fractions and mixed numbers by writing out steps.

Mandy uses $\frac{3}{4}$ yard of ribbon to wrap each small present and $1\frac{2}{4}$ yards of ribbon to wrap each medium-sized present. She will wrap 3 small presents and 4 medium-sized presents. How many yards of ribbon will Mandy use?

Step 1: Multiply to find out the amount of ribbon used to wrap the small presents and the amount used to wrap the medium-sized presents.

$3 \times \frac{3}{4} = \frac{9}{4} = 2\frac{1}{4}$

She used $2\frac{1}{4}$ yards of ribbon to wrap the small presents and 6 yards of ribbon to wrap the medium-sized presents.

$4 \times 1\frac{2}{4} = 4 \times \frac{6}{4} = \frac{24}{4} = 6$

Step 2: Add to find the amount of ribbon used to wrap all of the presents.

$2\frac{1}{4} + 6 = 8\frac{1}{4}$ Mandy used $8\frac{1}{4}$ yards of ribbon to wrap all of the presents.

Solve. Use equations and a representation to show your work.

1. On Monday through Thursday of each week, Olivia runs $2\frac{3}{5}$ miles each day. On Friday through Sunday, she runs $3\frac{2}{5}$ miles each day.

From Monday through Thursday, Olivia runs _____ miles.

From Friday through Sunday, Olivia runs _____ miles.

How many miles does Olivia run in one 7-day week?

Solve Problems Involving Fractions and Mixed Numbers

Name ______________________________

Write a word problem that involves multiplication of fractions and mixed numbers using the given topic. Use equations and a drawing to show your work.

1. homework

2. exercise

Understand Tenths and Hundredths

Name ______________________________

Review

You can use representations to understand tenths and hundredths.

How can you represent 70 cents as a fraction of a dollar?

$\frac{7}{10}$ of a dollar

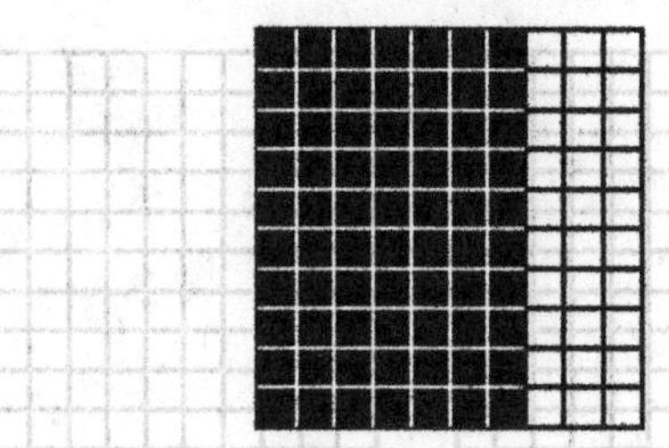

$\frac{70}{100}$ of a dollar

Which fraction does the grid represent?

1. 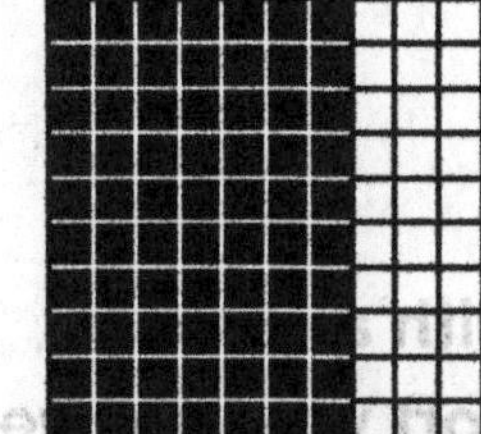 ______ ______

2.  ______ ______

How can you express the fraction as an equivalent fraction with a denominator of 10 or 100? Complete the equation.

3. $\frac{40}{100} = \frac{\square}{10}$

4. $\frac{17}{10} = \frac{\square}{100}$

5. $\frac{70}{100} = \frac{\square}{10}$

6. $\frac{3}{10} = \frac{\square}{100}$

Understand Tenths and Hundredths

Name ______________________________

Write each mixed number as an improper fraction with a denominator of 100. Shade the decimal grids to support your answer.

1. 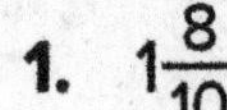 $1\frac{8}{10}$

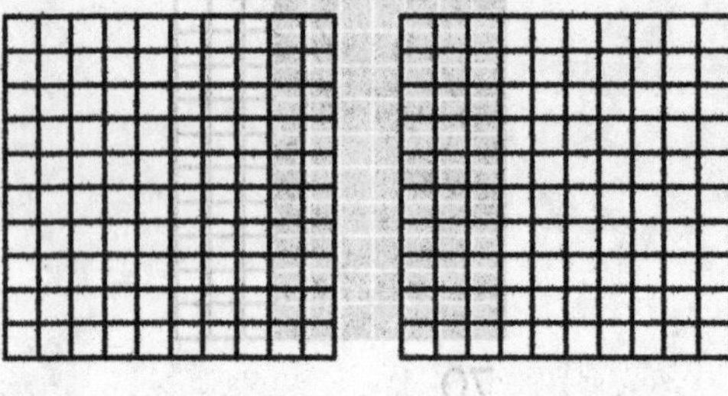

2. $2\frac{6}{10}$

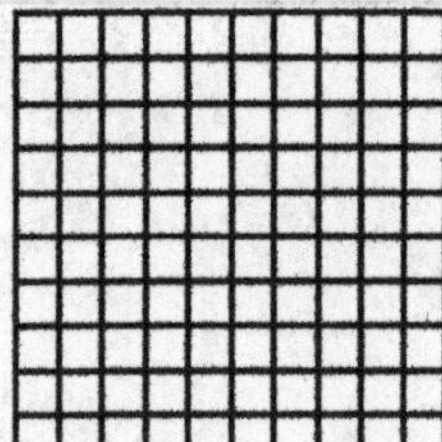 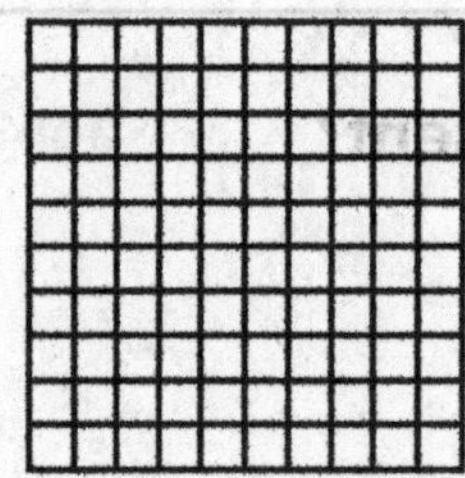 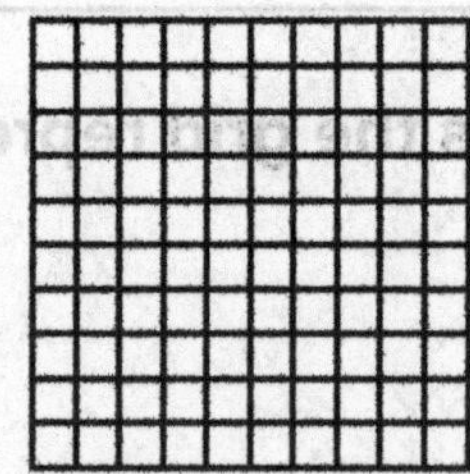

Write each mixed number as an improper fraction with a denominator of 10. Shade the decimal grids to support your answer.

3. $1\frac{40}{100}$

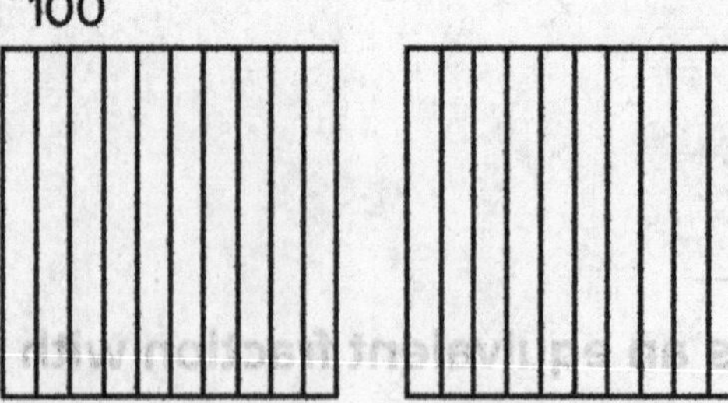

4. 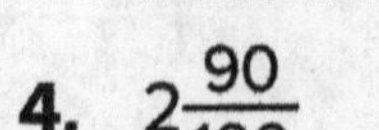 $2\frac{90}{100}$

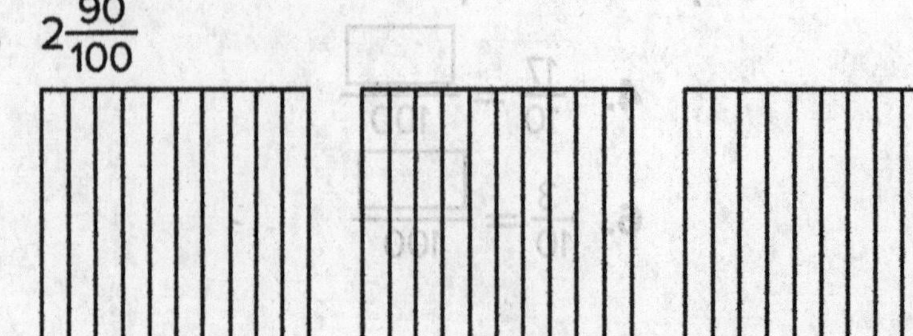

Understand Decimal Notation

Name ______________________________

Review

You can represent decimal fractions using decimal notation.

$\frac{3}{10}$ can be represented by the decimal grids and decimals shown below.

ones	tenths
0	3

ones	tenths	hundredths
0	3	0

Both 0.3 and 0.30 represent three tenths.

What decimal is represented by the model?

1.

ones	tenths	hundredths

2.

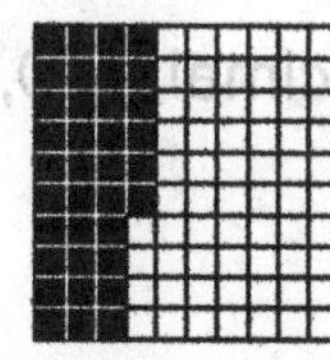

ones	tenths	hundredths

3.

ones	tenths	hundredths

4.

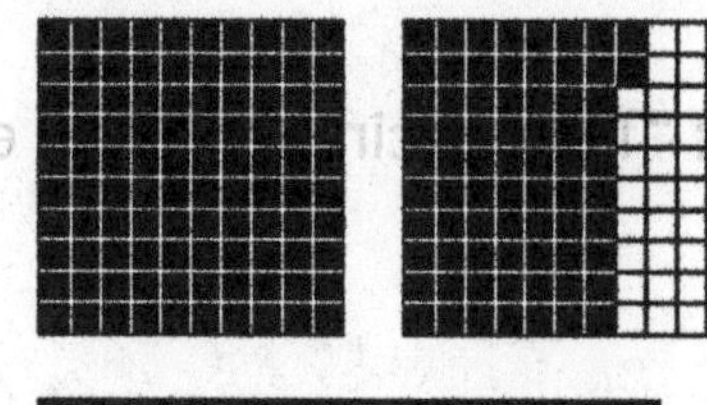

ones	tenths	hundredths

Understand Decimal Notation

Name ______________________________

Given below are statements made by different students. Explain any errors in their thinking. Use a model to support your response.

1. Randy: "The decimal, 1.2, is equivalent to $1\frac{20}{10}$."

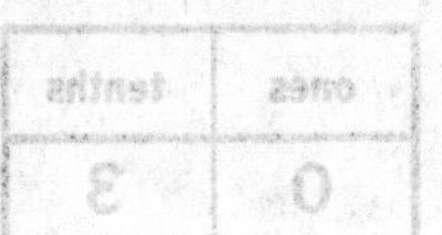

2. Edward: "The decimal, 2.3, is equivalent to $\frac{2}{3}$."

3. Christine: "The decimal, 1.70, is equivalent to $\frac{1}{7}$."

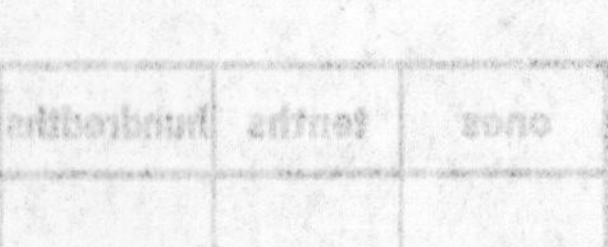

4. Susan: "The decimal, 0.9, is equivalent to $1\frac{9}{10}$."

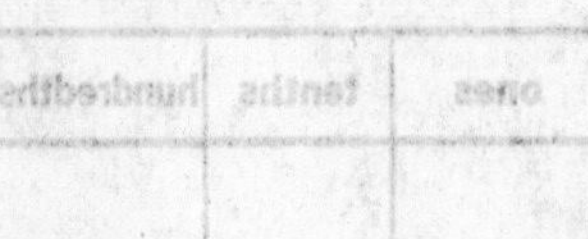

Compare Decimals

Name ______________________________

Review

You can use a number line to compare decimals.

0.6 and 0.45

0.6 > 0.45 since 0.6 falls to the right of 0.45 on the number line.

Which inequality makes the equation true, >, <, or = ? Complete the model to show the comparison.

1. 0.03 _____ 0.30

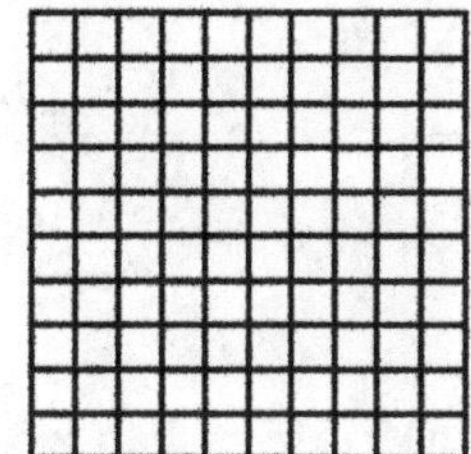

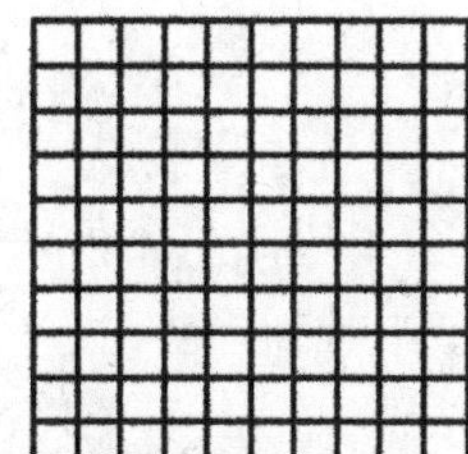

2. 0.7 _____ 0.70

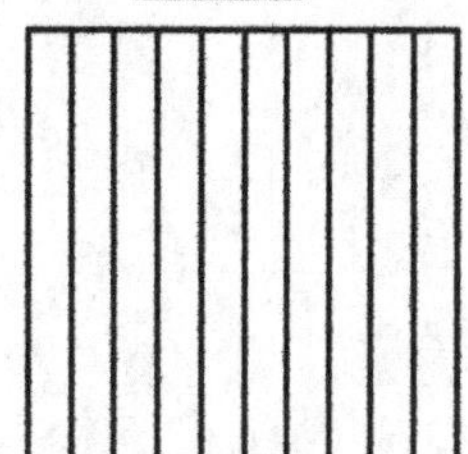

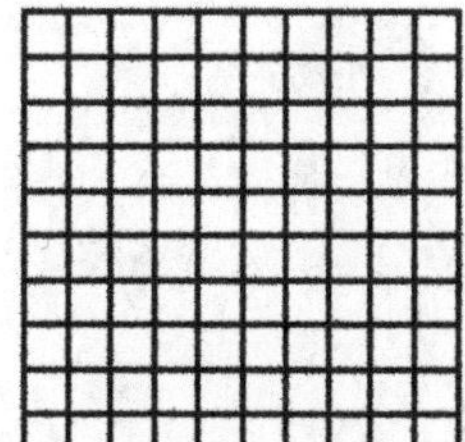

Complete the equations and inequality to compare the decimals.

3. 0.46 _____ 0.5

$0.46 = \frac{\quad}{100}$

$0.5 = \frac{\quad}{100}$

4. 0.43 _____ 0.34

$0.43 = \frac{\quad}{100}$

$0.34 = \frac{\quad}{100}$

Lesson **12-3 • Extend Thinking**

Compare Decimals

Name ______________________________

What missing decimal makes the inequality true? Use equivalent fractions and a number line to show your work.

1. $a > 1.4$

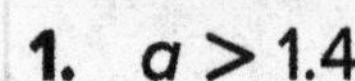

2. $a < 2.07$

3. $a = 1.9$

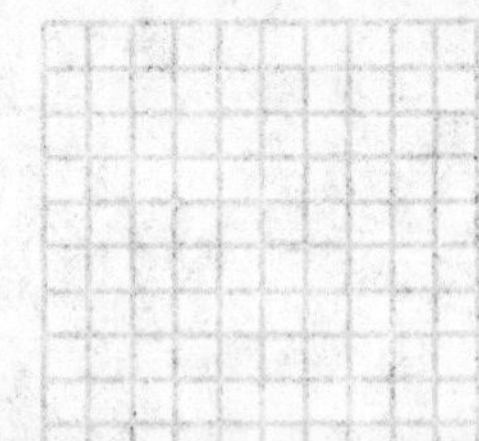

4. $a > 2.01$

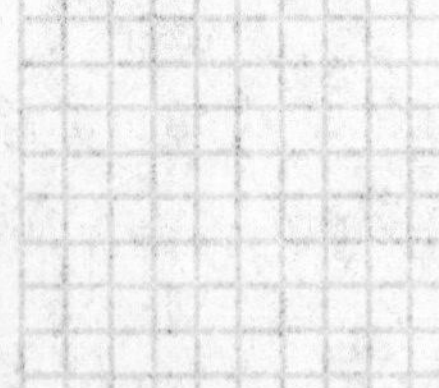

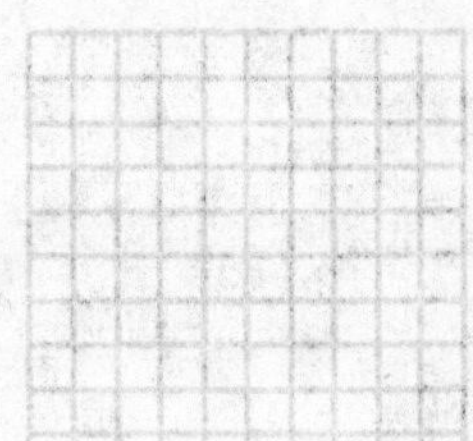

5. $a < 1.75$

6. $a = 2.60$

Adding Decimals Using Fractions

Name ______________________________

Review

You can use decimal grids to add decimal fractions.

$\frac{8}{10} + \frac{2}{100} = x$ *Represent $\frac{8}{10}$ as $\frac{80}{100}$*

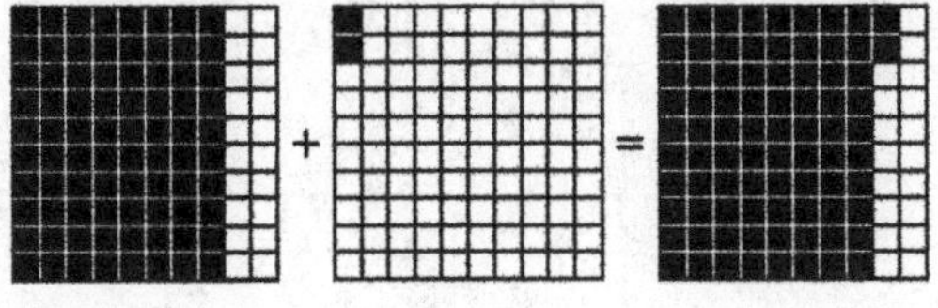

$\frac{8}{10} + \frac{2}{100} = \frac{82}{100}$

What is the sum of the fractions? Use the decimal grids to show your work.

1. $\frac{6}{10} + \frac{3}{100} =$ ____

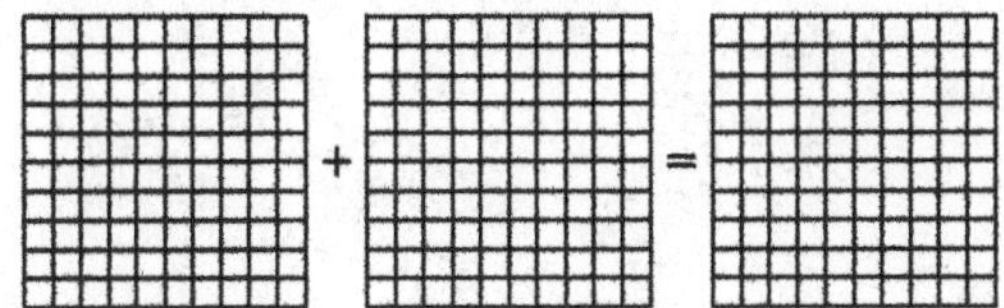

2. $\frac{32}{100} + \frac{4}{10} =$ ____

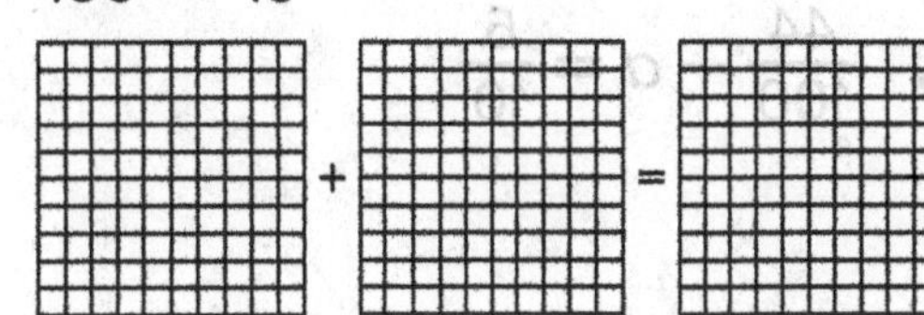

3. $\frac{7}{10} + \frac{11}{100} =$ ____

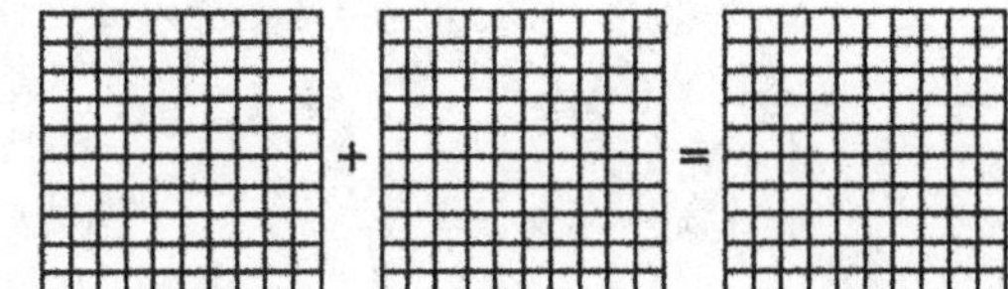

4. $\frac{50}{100} + \frac{5}{10} =$ ____

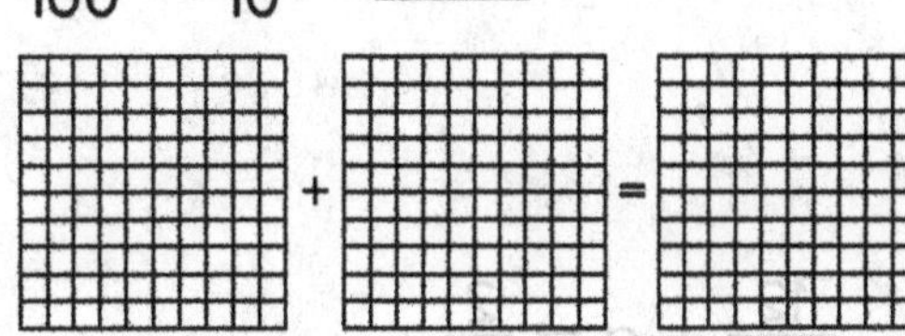

Adding Decimals Using Fractions

Name ______________________________

What missing fraction makes the equation true? Use an equation and a model to show your work.

1. $\frac{2}{10} + a = \frac{30}{100}$

2. $\frac{5}{100} + a = \frac{7}{10}$

3. $\frac{8}{10} + a = \frac{98}{100}$

4. $\frac{44}{100} + a = \frac{6}{10}$

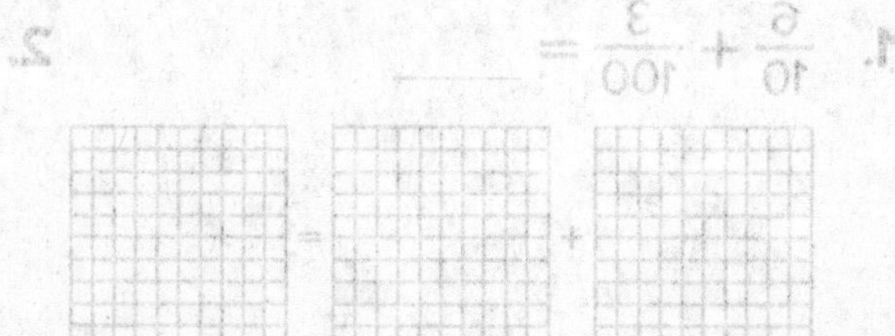

5. $\frac{1}{10} + a = \frac{23}{100}$

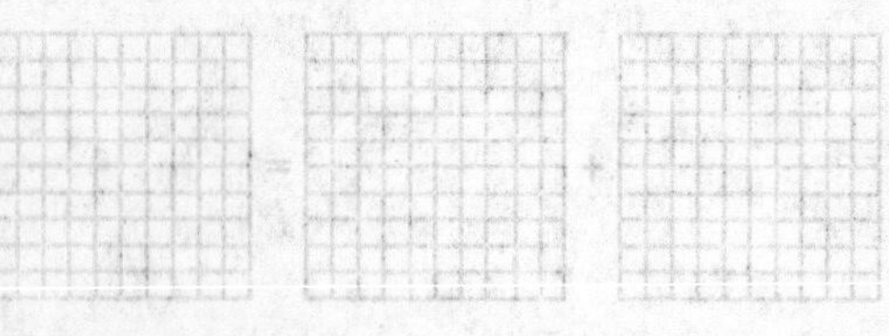

6. $\frac{12}{100} + a = \frac{2}{10}$

Lesson **12-5 • Reinforce Understanding**

Solve Problems Involving Money

Name ______________________________

Review

You can use bills and coins to add money amounts.

$7.40 + $2.55 = ?

Total value of the bills:

$9.00

Total value of each type of coin:

$0.90 $0.05

Use a place value chart to write the total amount as a decimal.

ones	tenths	hundredths
9	9	5

$7.40 + $2.55 = $9.95

What is the decimal that represents the total amount of money?

1.

$______

2.

$______

Solve Problems Involving Money

Name ______________________________

Write an addition word problem that involves the money amounts given. Use an equation and drawing of bills and coins to show your work.

1. $3.74 $2.37

2. $4.29 $3.62

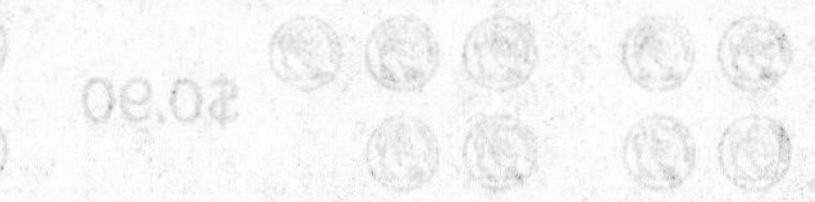

Write a subtraction word problem that involves the money amounts given. Use an equation and drawing of bills and coins to show your work.

3. $5.00 $2.35

4. $2.25 $1.72

Relate Metric Units

Name ______________________________

Review

You can multiply by 10s to convert larger units to smaller units.

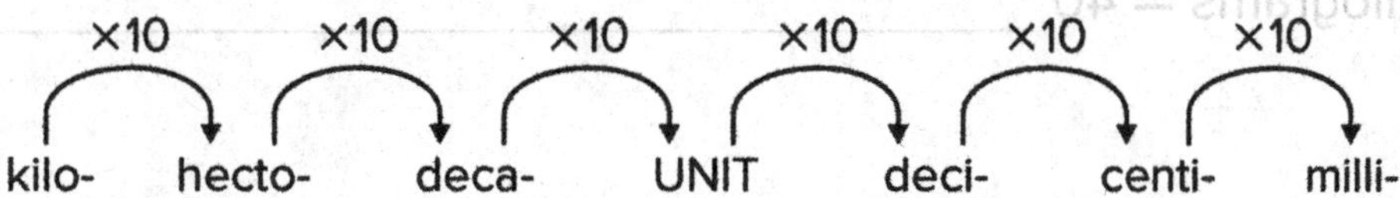

8 hectometers = ? decimeters

8 hectometers = 8 × 10 × 10 × 10 decimeters

= 8,000 decimeters

How can you convert the metric unit? Complete the equation using multiplication by 10s.

1. 6 meters = ? centimeters

6 meters = 6 × ______________ centimeters

6 meters = ______________ centimeters

2. 11 kilograms = ? grams

11 kilograms = 11 × ______________ grams

11 kilograms = ______________ grams

3. 23 centimeters = ? millimeters

23 centimeters = 23 × ______________ millimeters

23 centimeters = ______________ millimeters

4. 42 liters = ? deciliters

42 liters = 42 × ______________ deciliters

42 liters = ______________ deciliters

5. 3 decagrams = ? centigrams

3 decagrams = 3 × ______________ centigrams

3 decagrams = ______________ centigrams

Lesson **13-1 • Extend Thinking**

Relate Metric Units

Name ______________________________

What is the missing unit? Write the metric unit that makes the equation true.

1. 4 kilograms = 40 ______________________

2. 7 decagrams = 700 ______________________

3. 12 liters = 12,000 ______________________

4. 9 hectoliters = 90,000 ______________________

5. 150 decigrams = 1,500 ______________________

6. 22 deciliters = 2,200 ______________________

7. 75 kilograms = 750 ______________________

8. 100 grams = 100,000 ______________________

9. 18 kilometers = 18,000 ______________________

10. 25 grams = 2,500 ______________________

Relate Customary Units of Weight

Name ______________________________

Review

You can use multiplication to convert larger units to smaller units.

×2,000 ×16

Ton (T) → pound (lb) → ounce (oz)

How many ounces are in 6 pounds?

6 pounds = ? ounces

6 pounds = 6 × 16 ounces

6 pounds = 96 ounces

How many pounds are in a half ton?

$\frac{1}{2}$ ton = ? pounds

$\frac{1}{2}$ ton = $\frac{1}{2}$ × 2,000 pounds

$\frac{1}{2}$ ton = 1,000 pounds

Complete the equations to show how you can convert the customary units using multiplication?

1. 3 T = ? lb

3 T = 3 × __________ lb

3 T = __________ lb

2. 12 lb = ? oz

12 lb = 12 × __________ oz

12 lb = __________ oz

3. How many ounces are in $1\frac{1}{2}$ pounds?

4. How many pounds are in a quarter ton?

5. Carla's dog weighs $15\frac{1}{2}$ pounds. What is the weight of Carla's dog, in ounces?

Lesson **13-2 • Extend Thinking**

Relate Customary Units of Weight

Name ______________________________

Write a word problem that involves relating the customary units of weight. Solve. Explain how you arrived at your answer.

1. 2 tons = ? pounds

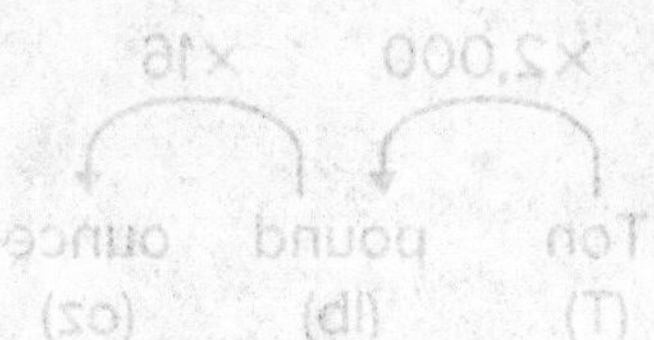

2. 2 pounds = ? ounces

3. $\frac{1}{4}$ ton = ? ounces

4. How can you use an equivalency table to convert $1\frac{1}{2}$ tons to pounds and ounces? Make an equivalency table to show your work.

Relate Customary Units of Capacity

Name ____________

Review

You can use multiplication to convert larger units to smaller units.

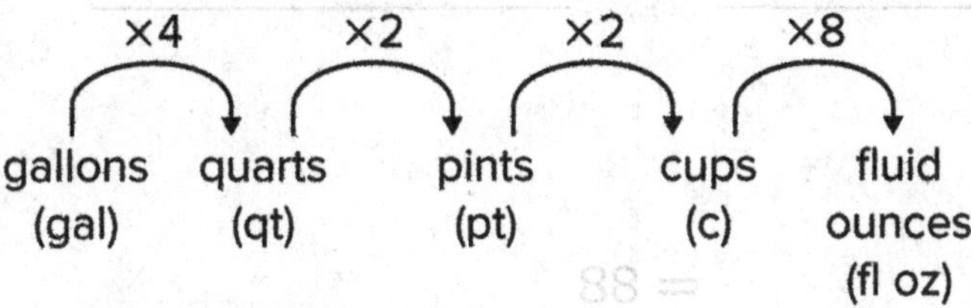

How many pints are in 6 gallons?

6 gallons = ? pints

6 gallons = 6 × 4 × 2 pints

6 gallons = 48 pints

There are 48 pints in 6 gallons.

Complete the equations to show how you can use multiplication to convert the customary units.

1. 5 pt = ? fl oz

   5 pt = 5 × ________ fl oz

   5 pt = ________ fl oz

2. 7 qt = ? c

   7 qt = 7 × ________ c

   7 qt = ________ c

3. How many fluid ounces are in 9 cups?

4. How many cups are in $1\frac{1}{2}$ pints?

5. Ron has $3\frac{1}{2}$ gallons of water. How many cups of water does Ron have?

Lesson **13-3 • Extend Thinking**

Relate Customary Units of Capacity

Name ______________________________

What two units of capacity make the equation true. Explain your thinking.

1. 4 ________________ = 64 ________________

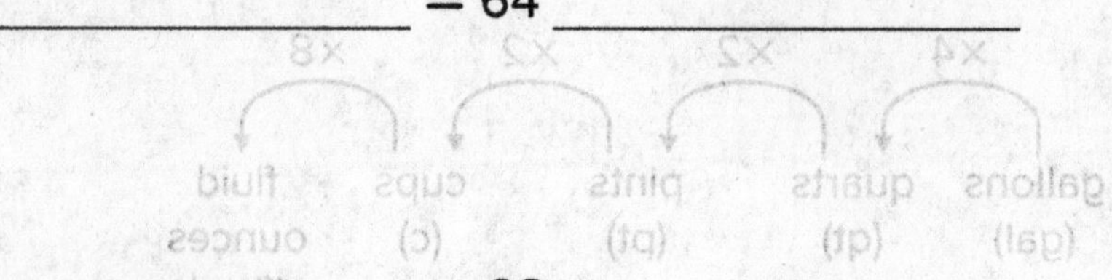

2. 11 ________________ = 88 ________________

3. 8 ________________ = 128 ________________

4. 10 ________________ = 40 ________________

5. 3 ________________ = 384 ________________

6. 10 ________________ = 320 ________________

Units of Time

Name ______________________________

Review

You can use multiplication to convert larger units to smaller units.

days →(×24)→ hours →(×60)→ minutes →(×60)→ seconds

$1\frac{1}{3}$ days = ? minutes

$1\frac{1}{3}$ days $= 1\frac{1}{3} \times 24 \times 60$ minutes

$1\frac{1}{3}$ days $= 1\frac{1}{3} \times 1{,}440$ minutes

$1\frac{1}{3}$ days $= 1 \times 1{,}440 + \frac{1}{3} \times 1{,}440$ minutes

$1\frac{1}{3}$ days $= 1{,}440 + 480$ minutes

$1\frac{1}{3}$ days $= 1{,}920$ minutes

How you can use multiplication to convert the units of time? Find the missing values.

1. 7 hr = ? min

7 hr = 7 × __________ min

7 hr = __________ min

2. $1\frac{1}{2}$ hr = ? sec

$1\frac{1}{2}$ hr = 1.5 × ______________ sec

$1\frac{1}{2}$ hr = ______________ sec

3. How many hours are in 3 days?

4. How many minutes are in $10\frac{1}{3}$ hours?

5. Kim spends 3 hours, 20 minutes in the kitchen. How many total minutes does Kim spend in the kitchen?

Units of Time

Name ______________________________

What two units of time make the equation true? Choose from the units: "*days*", "*hours*", "*minutes*" and "*seconds*". Explain your thinking.

1. 2 ____________ = 120 ____________

2. $4\frac{1}{3}$ ____________ = 104 ____________

3. 5 ____________ = 7,200 ____________

4. 10 ____________ = 1,680 ____________

5. $1\frac{1}{2}$ ____________ = $10\frac{1}{2}$ ____________

6. $\frac{1}{2}$ ____________ = 1,800 ____________

Solve Problems That Involve Units of Measure

Name ______________________________

Review

You can use equations and a bar diagram to represent a problem involving units of measure.

Cody has 2 liters and 250 milliliters of tea. He needs 3,000 milliliters of tea. How many more milliliters of tea will Cody have to make?

2 liters	250 milliliters	x milliliters

|- - - - - - - - 3,000 milliliters - - - - - - - -|

2 liters = 2 × 1,000 milliliters

2 liters = 2,000 milliliters

$2{,}000 + 250 + x = 3{,}000$

$2{,}250 + x = 3{,}000$

$x = 750$

Cody will have to make 750 milliliters more tea.

Eric can lift weights totaling 50 kilograms and 500 grams. He has a goal of lifting 75 kilograms. How many more grams of weight will Eric need to be able to lift to meet his goal?

1. Fill in the bar diagram with the missing values and units of measure.

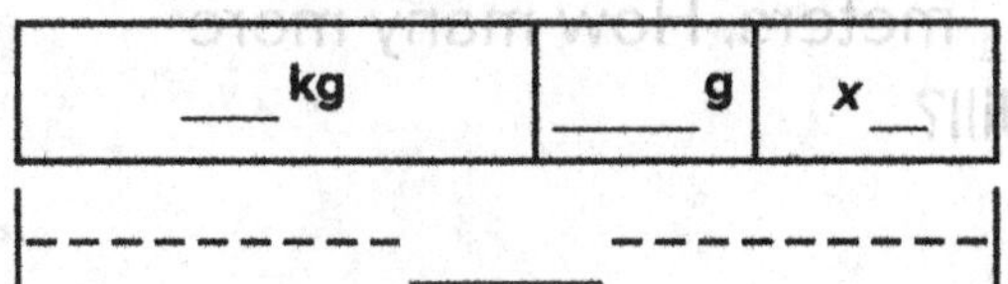

2. Fill in the bar diagram using grams as the unit of measure.

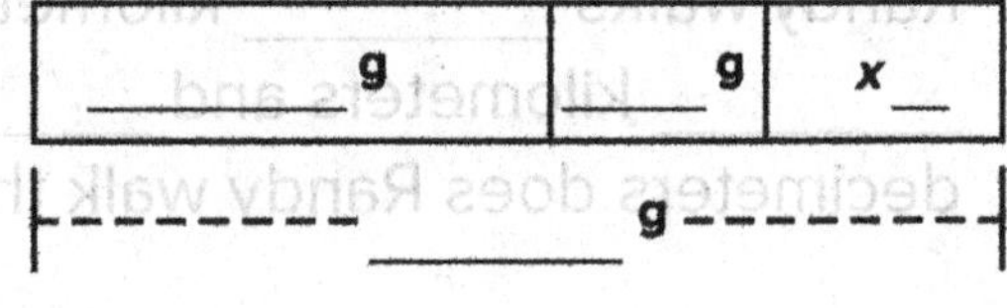

3. What equation can you write to solve the problem?

4. How many more grams will Eric need to be able to lift to meet his goal?

Solve Problems That Involve Units of Measure

Name ______________________________

What whole numbers completes the word problem? Create a bar diagram to represent the values you chose. Then, solve the problem and explain how you found your answer.

1. Ann has ________ quarts and ________ pints of lemonade. She needs ________ gallons of lemonade for a party. How many more cups of lemonade does Ann need?

2. Fred's grocery bag weighs ________ hectograms and ________ grams. Erica's grocery bag weighs ________ hectograms and ________ grams. How many more grams does Fred's grocery bag weigh than Erica's bag?

3. Randy walks ________ kilometers and ________ meters. Jill walks ________ kilometers and ________ meters. How many more decimeters does Randy walk than Jill?

Solve More Problems That Involve Units of Measure

Name ______________________________

Review

You can draw a clock to solve a problem that involves elapsed time.

Victor started mowing his lawn at 1:25 P.M. and finished at 3:40 P.M. How many minutes did it take Victor to mow his lawn?

1:25 + 1 hour = 2:25
2:25 + 1 hour = 3:25

3:25 + 1 hour = 4:25 X

3:25 + 5 minutes = 3:30
3:30 + 5 minutes = 3:35

3:35 + 5 minutes = 3:40
5 + 5 + 5 = 15 minutes

Victor spent 2 hr and 15 min

2 hr = 2 × 60 min = 120 min.

120 + 15 = 135

It took Victor 135 minutes to mow his lawn.

Solve. Use equations and a representation to show your work.

1. Allison left home at 11:45 A.M. She arrived home at 5:10 P.M the same day. How many minutes was Allison away from home?

Allison was away from home for _________ hours and _________ minutes.

How many minutes was Allison away from home?

Solve More Problems That Involve Units of Measure

Name ______________________________

What is the elapsed time for each movie? Complete the table.

Movie	Start	Stop	Elapsed Time (in minutes)
Movie A	2:15 P.M.	5:10 P.M.	
Movie B	9:35 A.M.	1:05 P.M.	
Movie C	12:40 P.M.	2:25 P.M.	

In the space below, explain how you arrived at each answer and use a number line to show your work.

Movie A:

Movie B:

Movie C:

Lesson **13-7 • Reinforce Understanding**

Solving Problems Using a Perimeter Formula

Name ______________________________

Review

You can use a formula to find the perimeter of a rectangle.

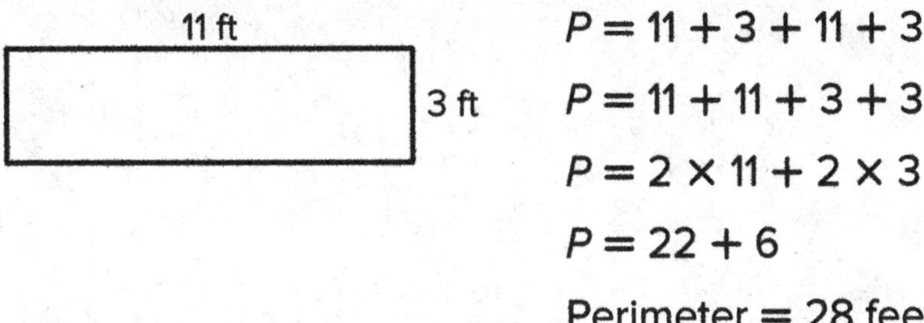

$P = 11 + 3 + 11 + 3$

$P = 11 + 11 + 3 + 3$

$P = 2 \times 11 + 2 \times 3$

$P = 22 + 6$

Perimeter = 28 feet

How can you find the perimeter of a rectangle using a formula? Complete the equations to find the perimeter of the rectangle with the given dimensions.

1. $l = 8$ in., $w = 3$ in.

$P = 2 \times$ ____ $+ 2 \times$ ____

$P =$ ____ $+$ ____

$P =$ ________

2. $l = 22$ cm, $w = 14$ cm

$P = 2 \times$ ____ $+ 2 \times$ ____

$P =$ ____ $+$ ____

$P =$ ________

3. $l = 5$ ft., $w = 7$ ft.

$P = 2 \times$ ____ $+ 2 \times$ ____

$P =$ ____ $+$ ____

$P =$ ________

4. $l = 25$ m, $w = 100$ m

$P = 2 \times$ ____ $+ 2 \times$ ____

$P =$ ____ $+$ ____

$P =$ ________

What is the length of the missing dimension?

5. A rectangular poster has a perimeter of 120 inches and a length of 36 inches. What is the width of the poster?

6. A rectangular piece of fabric on a lawn chair has a perimeter of 22 feet and a width of 2 feet. What is the length of the piece of fabric?

Solving Problems Using a Perimeter Formula

Name ______________________________

What are possible dimensions of the rectangle with the given perimeter? Explain how you arrived at your answer.

1. $P = 28$ cm

2. $P = 50$ in.

3. $P = 18$ ft

4. $P = 40$ in.

5. $P = 24$ yd

6. $P = 866$ cm

7. $P = 222$ cm

8. $P = 100$ in.

Solve Problems Using an Area Formula

Name ______________________________

Review

You can use a formula to find the area of a rectangle.

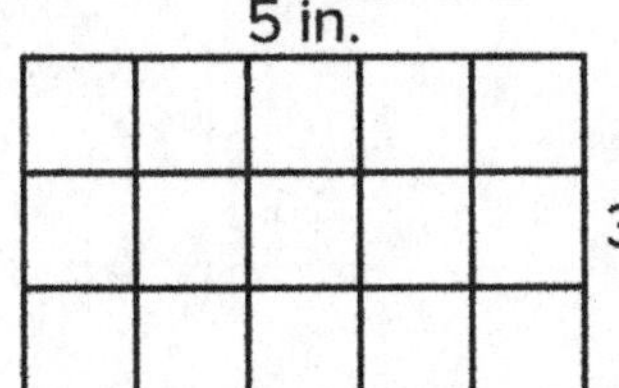

$A = l \times w$

$= 5 \times 3$

$= 15$

Area = 15 square inches

How can you find the area of a rectangle using a formula? Fill in the equations to show how.

1. $l = 11$ cm, $w = 9$ cm

   $A =$ _____ × _____

   $A =$ _____ square cm

2. $l = 7$ in., $w = 12$ in.

   $A =$ _____ × _____

   $A =$ _____ square in.

What is the missing measure? Make sure to include the units.

3. A rectangular cement block has a length of 14 inches and a width of 10 inches. What is the area of the cement block?

4. A square mouse pad has a side length of 7 inches. What is the area of the pad?

5. A rectangular lid has an area of 36 square cm and a length of 9 cm. What is the width of the lid?

6. A rectangular flower bed has an area of 36 square feet. What are 3 possible dimensions for the bed?

7. Prabah says that the area of a canvas with a length of 12 inches and width of 14 inches is 26 square inches. What mistake did Prabah make?

Solve Problems Using an Area Formula

Name ______________________________

What are possible dimensions for the rectangle with the given area? Explain how you found your answer.

1. $A = 72$ square feet

2. $A = 108$ square inches

3. $A = 56$ square centimeters

Which of the following rectangles could have an area of 144 square feet? Answer *Yes* or *No*. If you answer "*Yes*" give the dimensions of the possible rectangle.

4. Rectangle with perimeter of 50 feet. ______________

5. Rectangle with perimeter 60 feet. ______________

6. Square with perimeter 24 feet. ______________

7. Rectangle with side length 144 feet. ______________

8. Rectangle with perimeter 40 feet. ______________

Solve Problems Involving Perimeter and Area

Name ______________________________

Review

You can use formulas to solve problems involving perimeter and area.

Elizabeth wants to create a rectangular flower bed with an area of 72 square feet. The width of the flower bed will be 8 feet.

What will be the dimensions of the flower bed?

The dimensions will be 8 ft × 9 ft.

$$\begin{cases} A = l \times w \\ 72 = l \times 8 \\ l = 9 \text{ ft} \end{cases}$$

What will be the perimeter of the flower bed?

The perimeter will be 34 ft.

$$\begin{cases} P = 2 \times w + 2 \times l \\ P = 2 \times 8 + 2 \times 9 \\ P = 16 + 18 \\ P = 34 \text{ ft} \end{cases}$$

Complete the equations to find the unknown measurements.

1. Rectangle: top l ft; side 6 ft; $A = 48$ sq ft

$48 = l \times 6$

$l =$ _____ ft

$P = 2 \times$ _____ $+ 2 \times$ _____

$P =$ _____ $+$ _____

$P =$ _____ ft

2. Rectangle: top 5 in.; side w in.; $A = 20$ sq in.

$20 = 5 \times w$

$w =$ _____ in.

$P = 2 \times$ _____ $+ 2 \times$ _____

$P =$ _____ $+$ _____

$P =$ _____ in

3. A gardener has 60 feet of fencing to fence a rectangular garden. The garden needs to be 10 feet long. What will be the area of the garden?

Solve Problems Involving Perimeter and Area

Name ______________________________

What could be the area of a rectangle with the given perimeter? Explain how you arrived at your answer.

1. $P = 46$ cm

2. $P = 30$ ft

The perimeter of a rectangle is 72 cm. What could be the area of this rectangle if the following changes are made? Explain how you arrived at your answer.

3. The width of the rectangle is doubled.

4. The width of the rectangle is increased by 2.

Display and Interpret Data on a Line Plot

Name ______________________________

Review

You can use a line plot to display and interpret data.

The line plot shows that:

- $5\frac{1}{4}$ in. was the most common pencil length. It has the most x's.
- None of the pencils were exactly 5 in. long, because there are no x's above the 5.
- The pencils were measured to the nearest $\frac{1}{4}$ in., because the number line has increments of $\frac{1}{4}$ in.
- 26 pencils were measured, because there are 26 x's.

4 $4\frac{1}{4}$ $4\frac{2}{4}$ $4\frac{3}{4}$ 5 $5\frac{1}{4}$ $5\frac{2}{4}$ $5\frac{3}{4}$ 6

Pencil Lengths (in.)

Use the line plot to answer the questions.

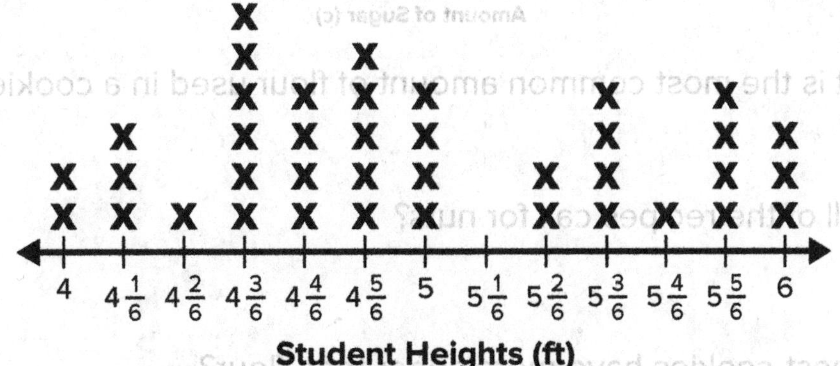

1. How many students were 4 ft tall?
2. What is the most common student height?
3. How many students are represented by the line plot?
4. How many students were $5\frac{1}{6}$ feet tall?
5. Student heights are measured to the nearest ______________

Display and Interpret Data on a Line Plot

Name ______________________________

The line plots below represent the amounts of ingredients included in 20 different cookie recipes. Use the line lots to answer the questions. Explain your thinking.

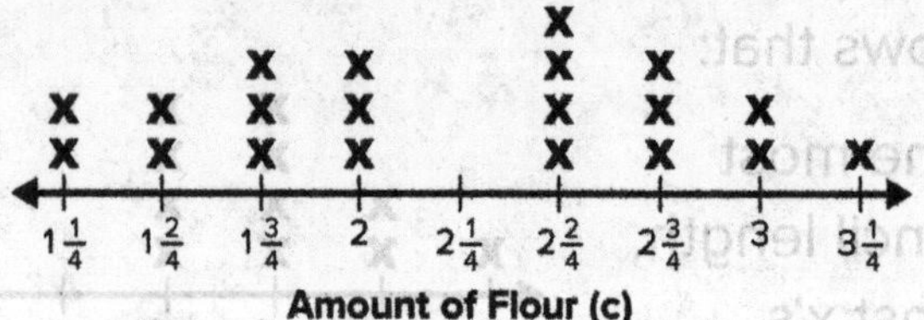

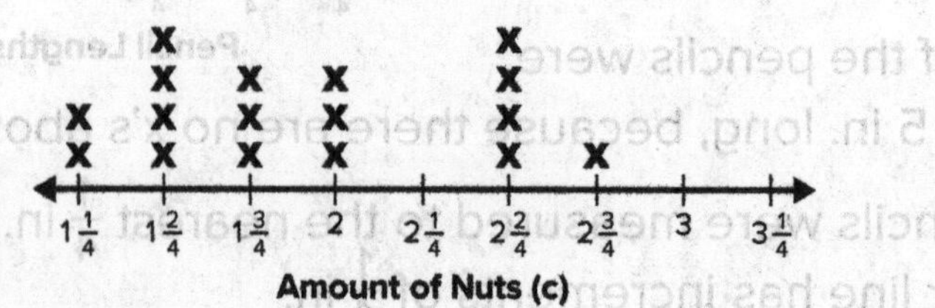

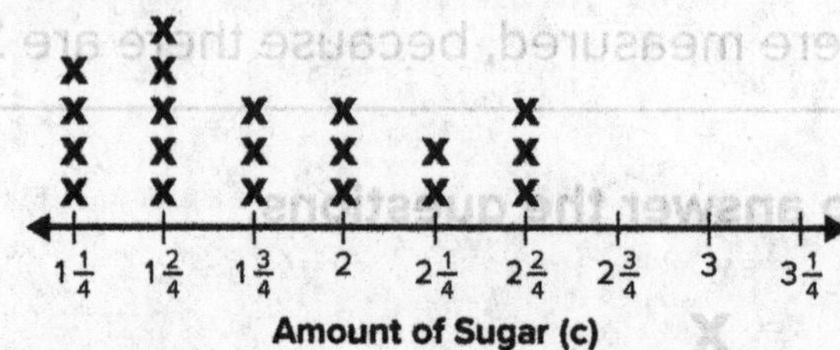

1. What is the most common amount of flour used in a cookie recipe?

2. Do all of the recipes call for nuts?

3. Do most cookies have more sugar than flour?

4. Is it possible that one of the recipes calls for more nuts than sugar?

Lesson **13-11 • Reinforce Understanding**

Solve Problems Involving Data on a Line Plot

Name ______________________________

Review

You can use a line plot to analyze measurement data.

What is the difference between the most common fruit weight and the second-most common fruit weight? $1\frac{2}{4} - \frac{3}{4} = \frac{3}{4}$.

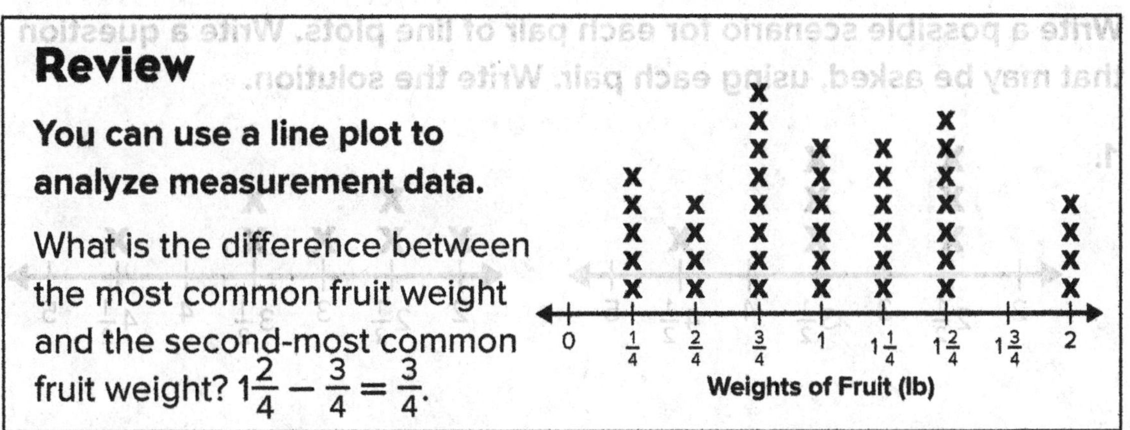

Use the 2 line plots to answer the question. Show your work.

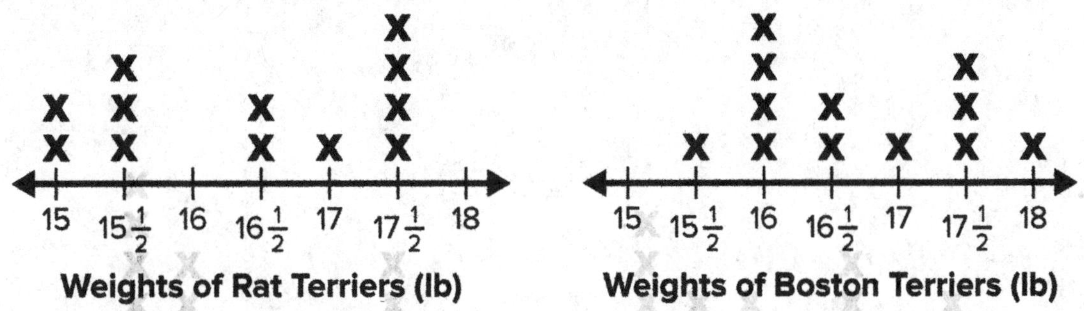

1. What is the difference between the most common Rat Terrier weight, and the most common Boston Terrier Weight?

3. What is the combined weight of the Boston Terriers?

4. What is the difference between the most common Boston Terrier weight and the second-most common Boston Terrier weight?

Solve Problems Involving Data on a Line Plot

Name ______________________________

Write a possible scenario for each pair of line plots. Write a question that may be asked, using each pair. Write the solution.

1.

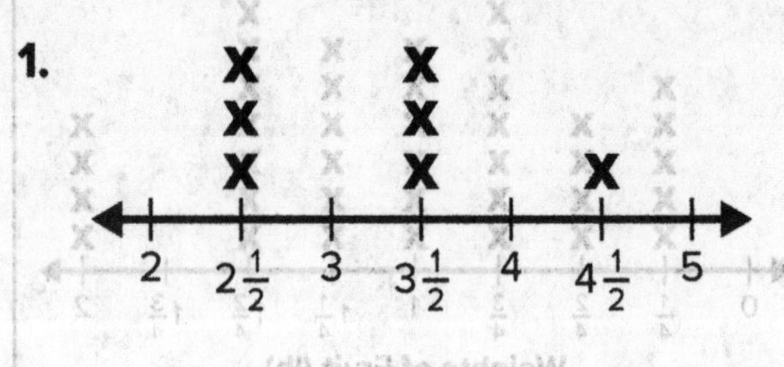

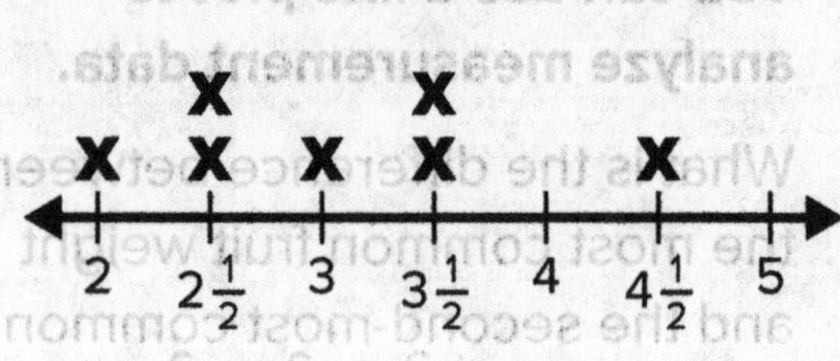

2.

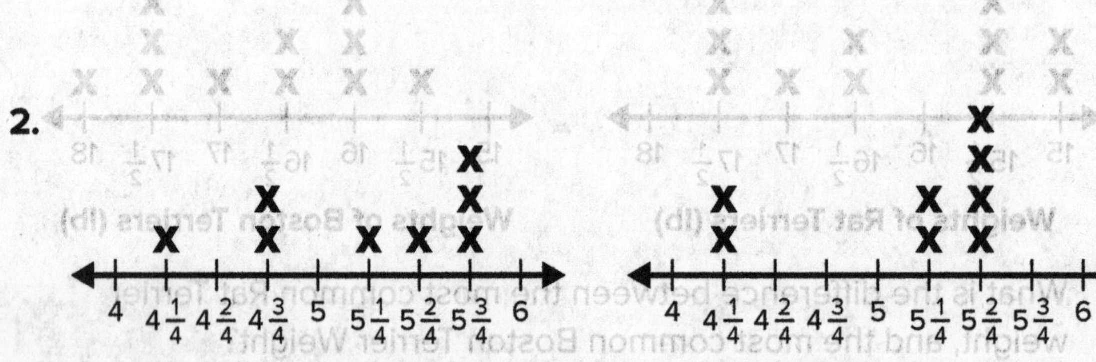

Understand Lines, Line Segments, and Rays

Name ______________________________

Review

You can use images of geometric figures to learn about their properties.

Point	A Point A	Location in space
Line	B A Line AB or $\overleftrightarrow{AB}$	Straight 1-dimensional figure that continues in both directions forever
Line Segment	E F Segment EF or $\overline{EF}$	Section of a line and has two endpoints
Ray	L K Ray LK or $\overrightarrow{LK}$	1-dimensional figure with one endpoint that extends forever in the other direction

Match the name to the figure.

1. line

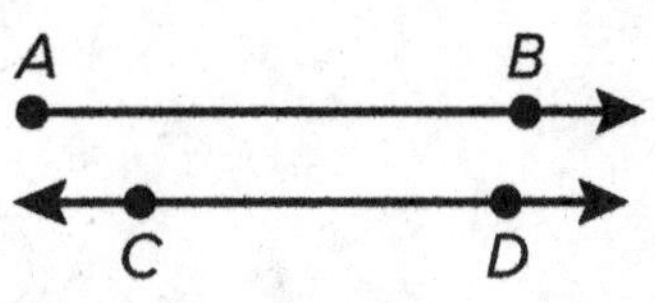

2. point

3. ray

4. line segment

P

Lesson **14-1 • Extend Thinking**

Understand Lines, Line Segments, and Rays

Name ______________________________

What is a real-world example that meets the specification? Then, provide a sketch of the example.

1. line segments only

2. line segments and rays

3. rays and lines

4. rays only

5. lines only

Classify Angles

Name ______________________________

Review

You can sort angles by using their properties.

<table>
<tr>
<td>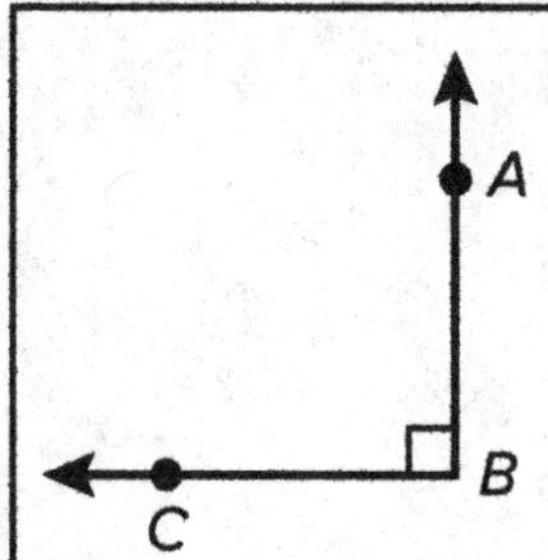</td>
<td>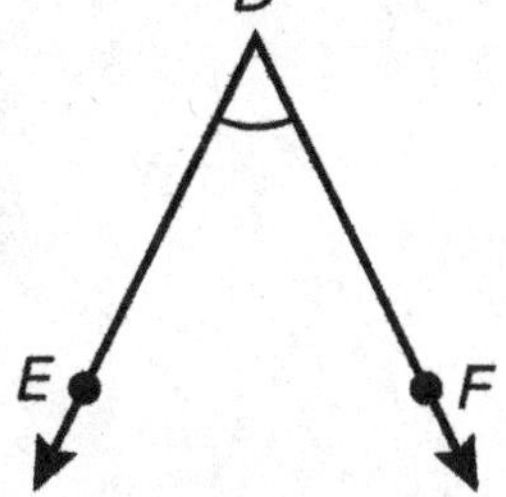</td>
<td>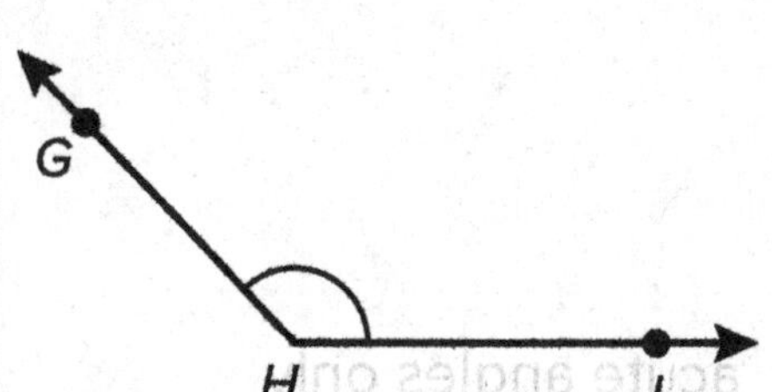</td>
</tr>
<tr>
<td>Rotation from one ray to the other is $\frac{1}{4}$ of the circle</td>
<td>Rotation from one ray to the other is less than $\frac{1}{4}$ of a circle</td>
<td>Rotation from one ray to the other is greater than $\frac{1}{4}$ of a circle</td>
</tr>
</table>

Match the name to the angle.

1. acute

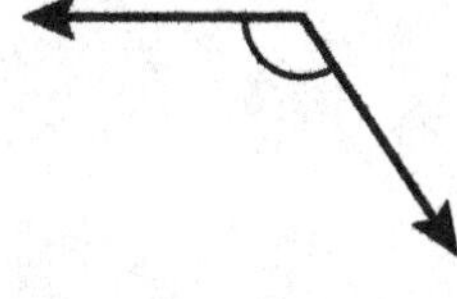

2. obtuse

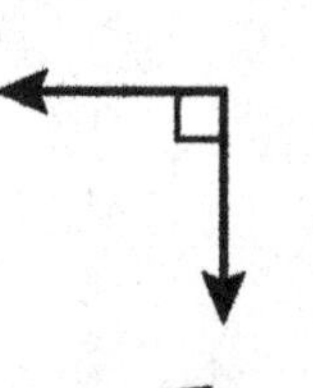

3. right

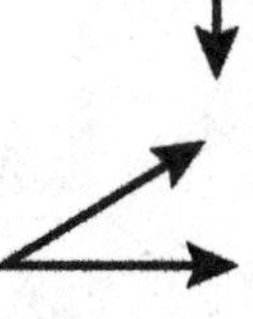

Name and draw a diagram of a real-world object that includes the specified angle type.

4. right **5.** acute **6.** obtuse

Classify Angles

Name ______________________________

What two-dimensional shape includes the type(s) of angles specified? Draw the shape.

1. obtuse angles only

2. acute angles only

3. right angles only

4. acute and obtuse angles

5. right, acute, and obtuse angles

Lesson **14-3 • Reinforce Understanding**

Draw and Measure Angles

Name ______________________________

Review

You can use a protractor to measure an angle.

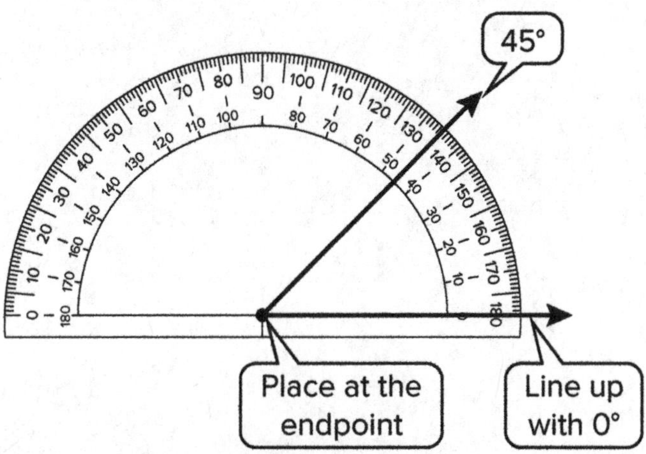

The tick mark that aligns with the second ray shows an angle measure of 45°.

What is the angle measure? Use a protractor. Extend the rays, if needed.

1. 

2. 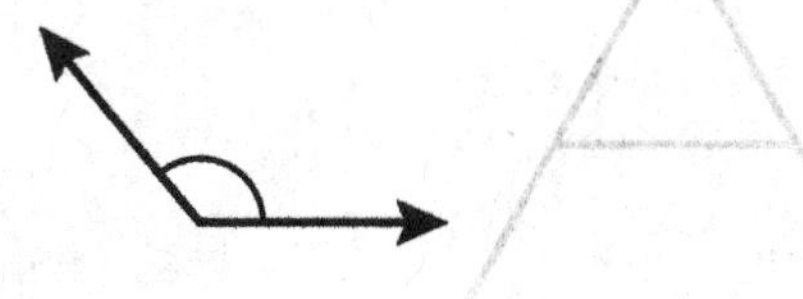

3. 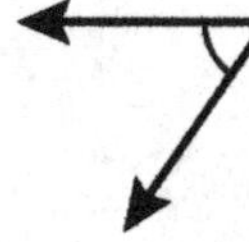

4. 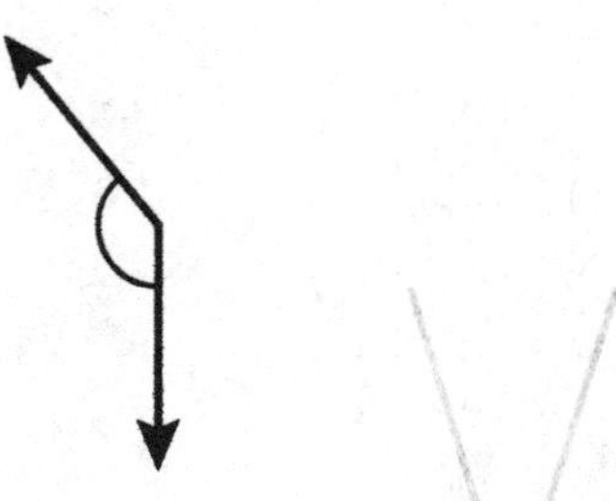

Use a protractor to draw the angle.

5. 135°

Draw and Measure Angles

Name ____________________

Measure the angles formed by the letter.

1.

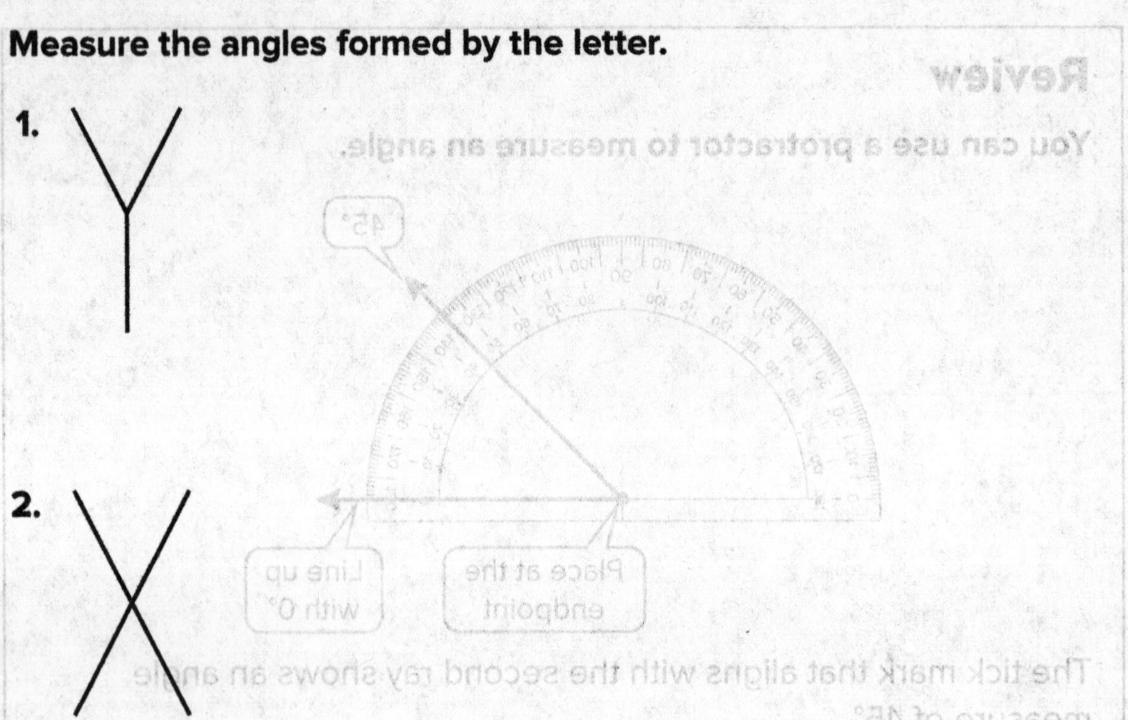

2.

3.

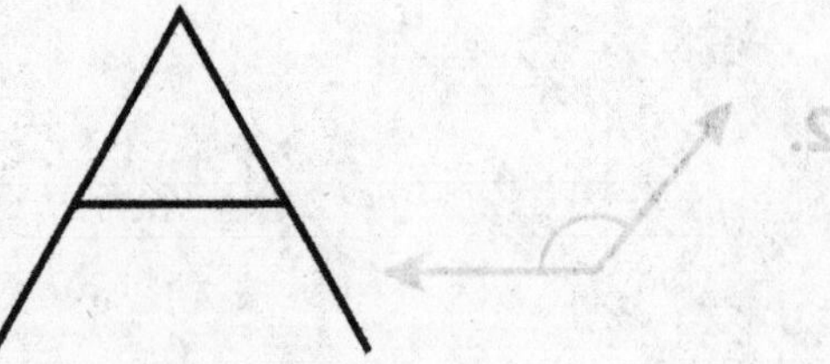

4.

Understand Parallel and Perpendicular Lines

Name ______________________________

Review

You can describe two paths according to whether they are parallel, perpendicular, or neither.

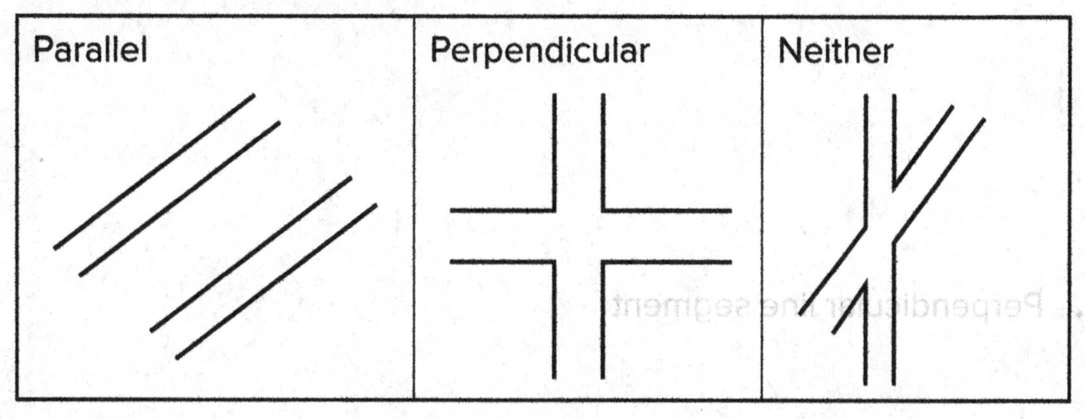

Draw a line, according to the given description.

1. parallel to

2. perpendicular to

3. neither parallel nor perpendicular to

4. List all of the capital letters that include parallel lines.

5. List all of the capital letters that include perpendicular lines.

Understand Parallel and Perpendicular Lines

Name ______________________________

What 2-dimensional figure contains the specification given? Draw the shape.

1. Parallel line segments

2. Perpendicular line segment

3. Parallel and perpendicular line segments

4. Neither parallel nor perpendicular line segments

Add and Subtract Angle Measures

Name ____________________

Review

You can use a protractor to find an unknown angle measure.

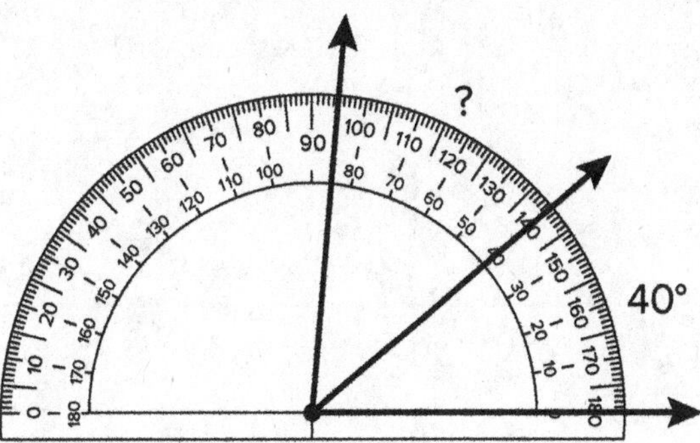

The other smaller angle measures 45°.

What is the sum of the two angles?

1.

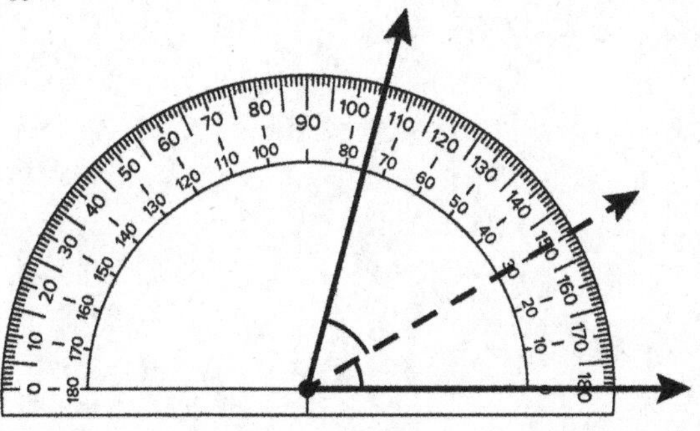

2.

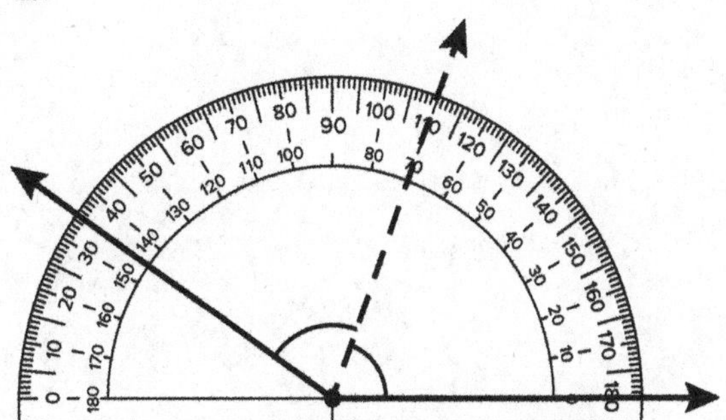

What is the measure of the unknown angle?

3.

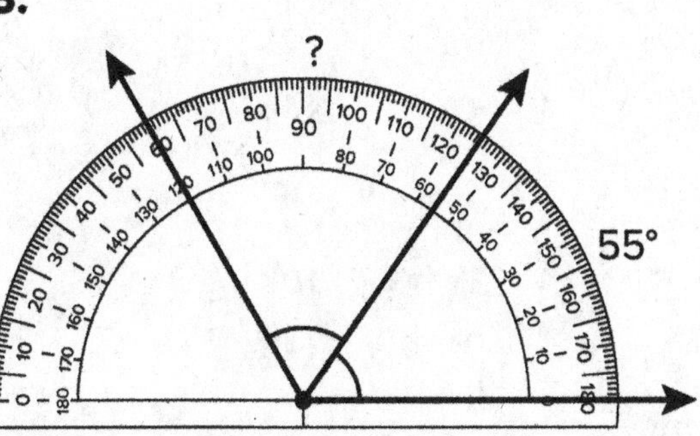

4.

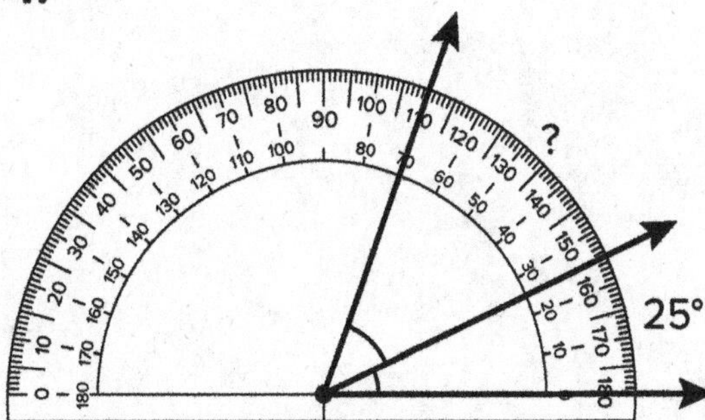

Lesson **14-5 • Extend Thinking**

Add and Subtract Angle Measures

Name ______________________________

What real-world object includes the type of angle given? Provide a sketch of the object and decompose the angle into two smaller angles. Measure the angles and label the angle measures.

1. acute angle

2. right angle

3. obtuse angle

Solve Problems Involving Unknown Angle Measures

Name ____________________

Review

You can use an equation to solve a problem involving angle measures.

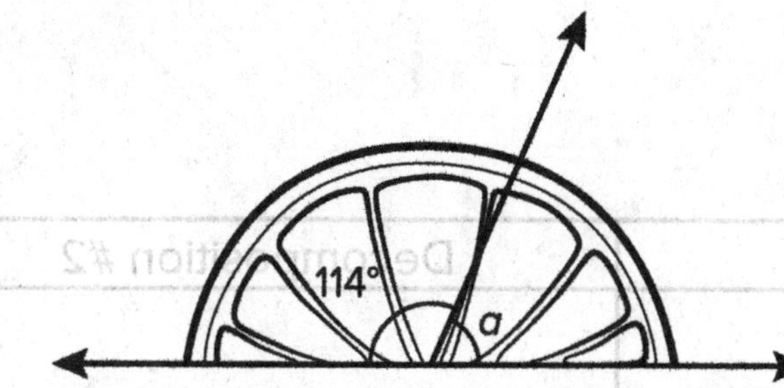

$a + 114° = 180°$

$180° - 114° = a$

$180° - 114° = 66°$

$a = 66°$

What is the combined angle measure? Write an equation to show your work.

1.

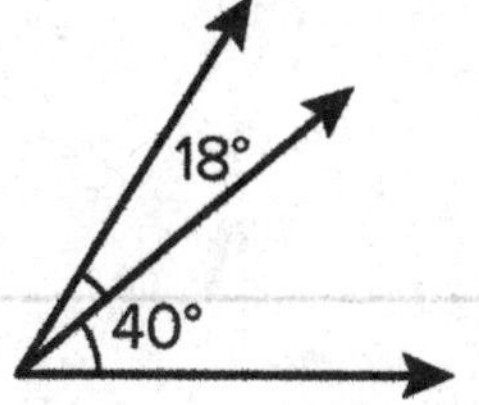

2.

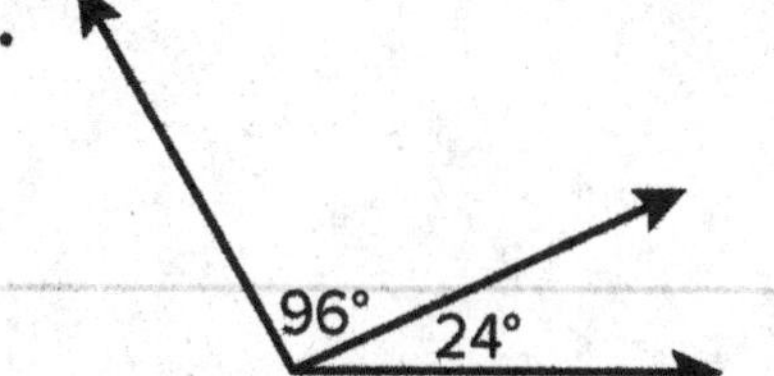

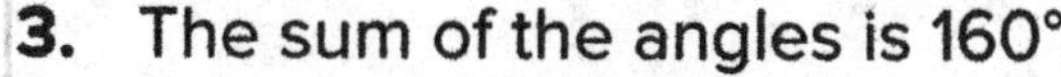

What is the unknown angle measure? Use an equation to show your work.

3. The sum of the angles is 160°.

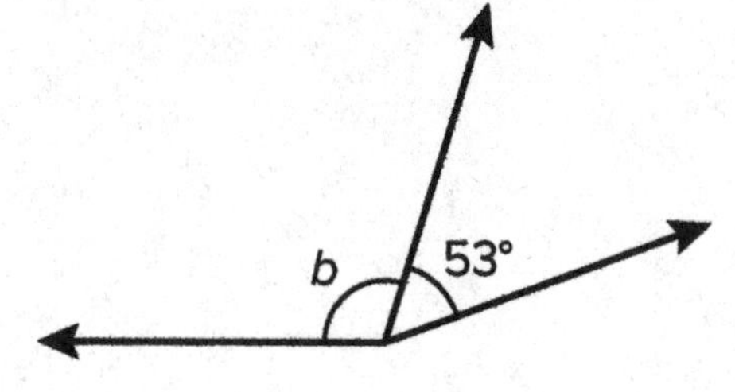

4. The sum of the angles is 57°.

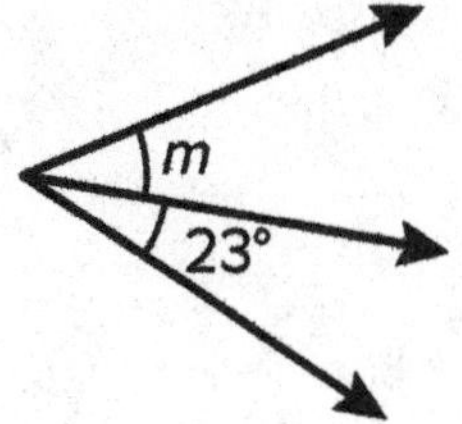

Solve Problems Involving Unknown Angle Measures

Name ______________________________

What are two different ways to decompose the angle into three smaller angles? For each decomposition, label the angle measures. Write an equation to represent the combined angle measure.

1.

Decomposition #1	Decomposition #2

2.

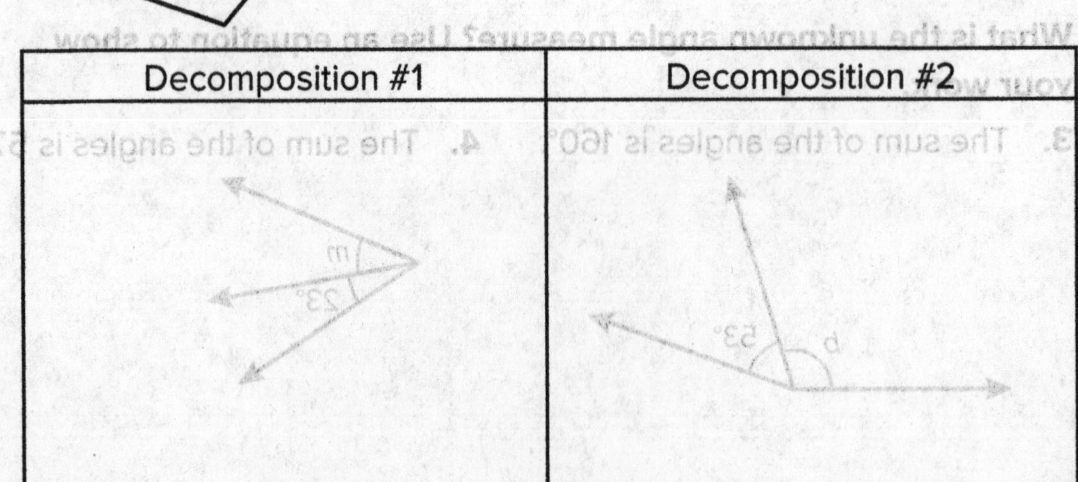

Decomposition #1	Decomposition #2

Classify Polygons

Name ________________________________

Review

You can use properties of two-dimensional figures to sort them.

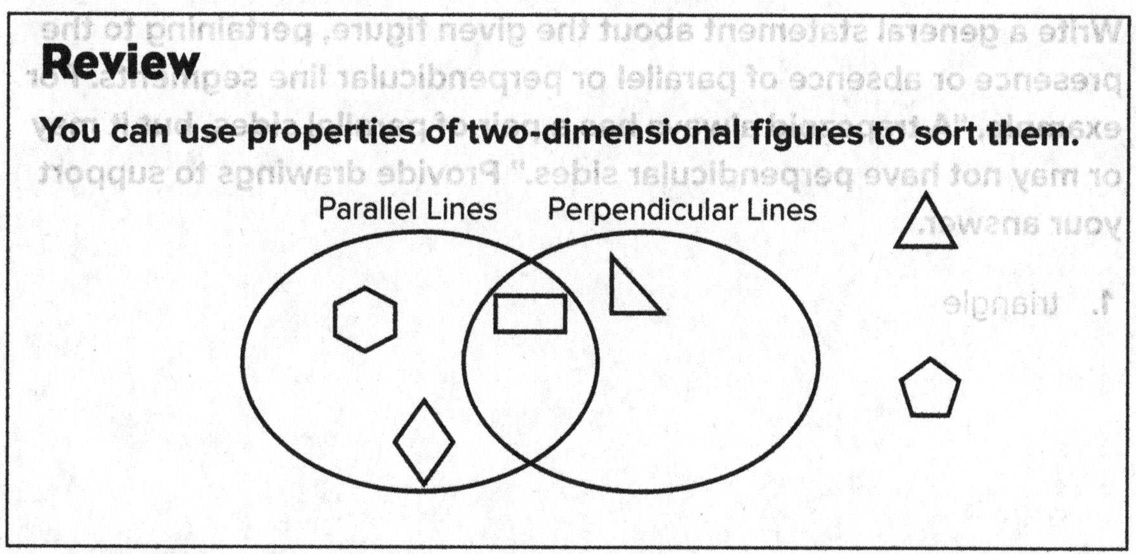

Which classification does the polygon fit into: parallel line segments only, perpendicular line segments only, both, or neither?

1. 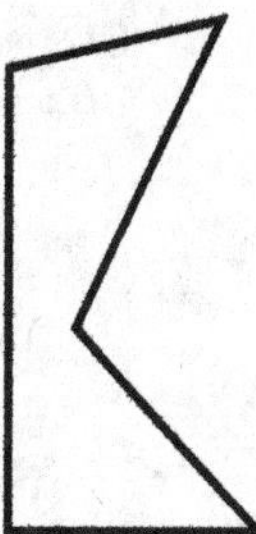

2. 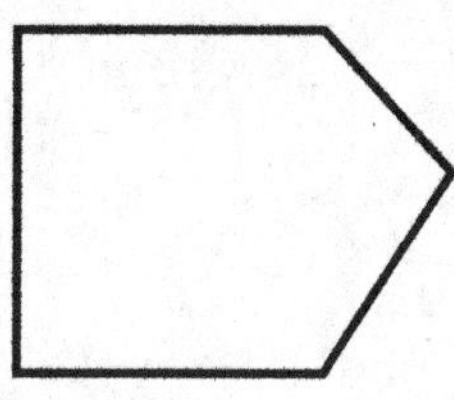

3. 

4. 

Classify Polygons

Name ____

Write a general statement about the given figure, pertaining to the presence or absence of parallel or perpendicular line segments. For example, "A trapezoid always has a pair of parallel sides, but it may or may not have perpendicular sides." Provide drawings to support your answer.

1. triangle

2. hexagon

3. pentagon

4. trapezoid

Classify Triangles

Name ______________________________

Review

You can sort triangles using their angles and side lengths.

Right Triangle

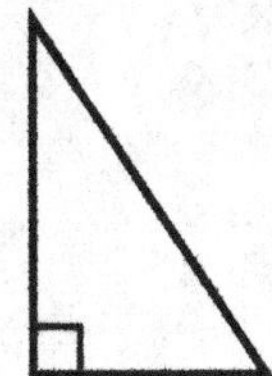

Acute Triangle

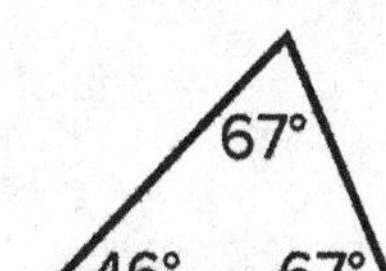

Obtuse Triangle

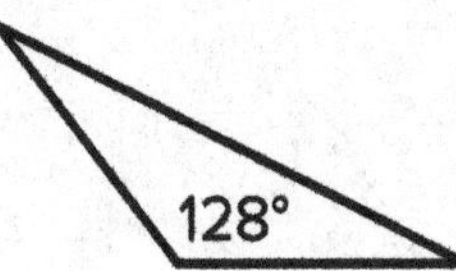

Equilateral Triangle

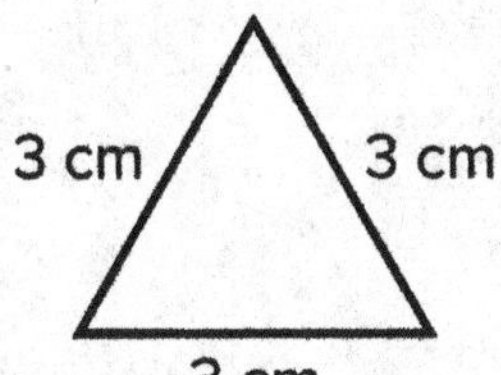

Isosceles Triangle

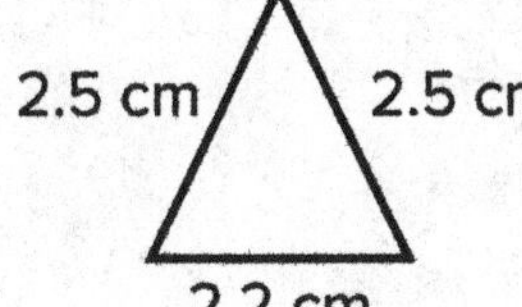

Scalene Triangle

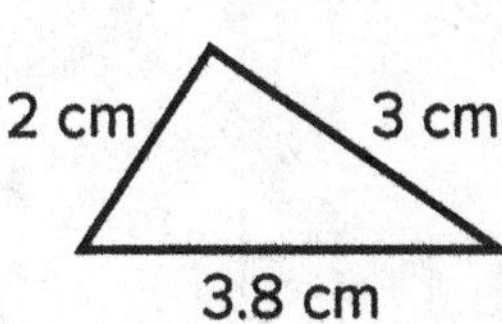

What is the classification of the triangle based on its angles?

1. 

2. 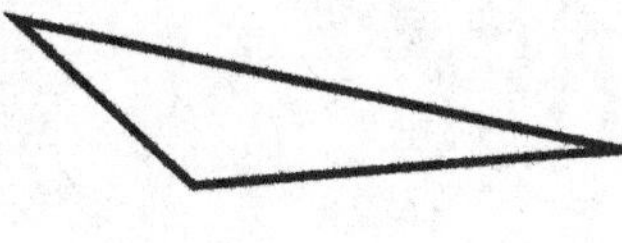

3. 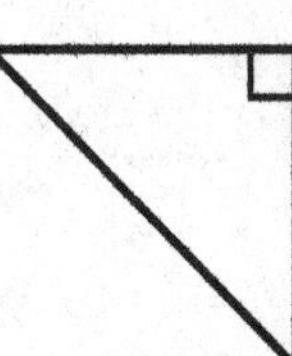

What is the classification of the triangle based on its angles?

4. 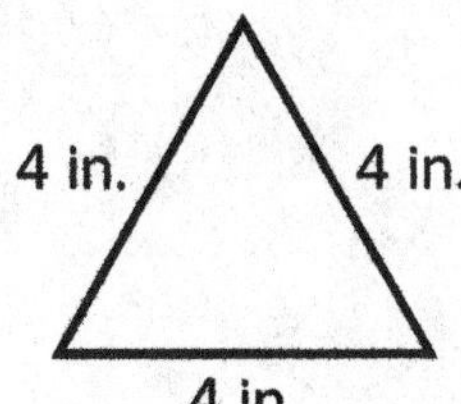

5. 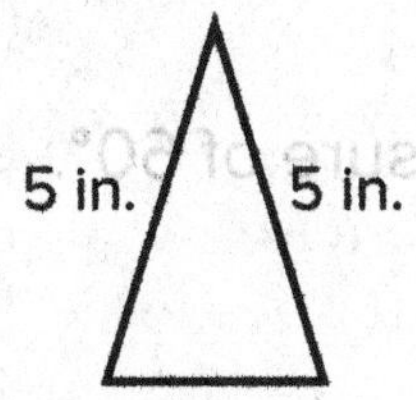

6. 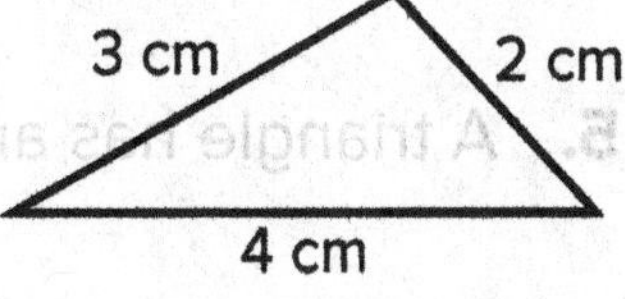

Lesson **14-8 • Extend Thinking**

Classify Triangles

Name ______________________________

Classifying by angles only, use the statement to decide which type of triangle is possible. List the possible triangles. Provide drawings to support your answer.

1. A triangle has an angle measure of 120°.

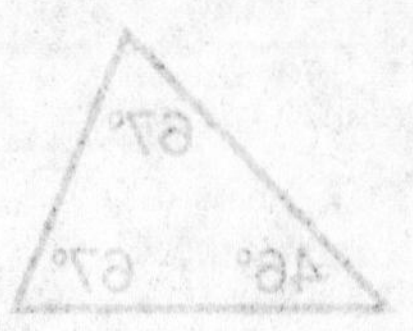

2. A triangle has an angle measure of 30°.

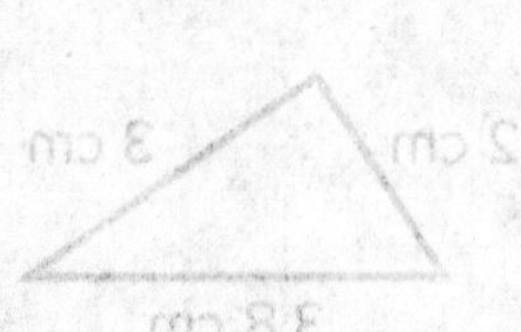
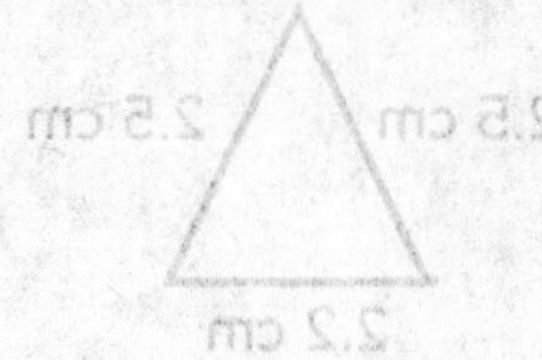
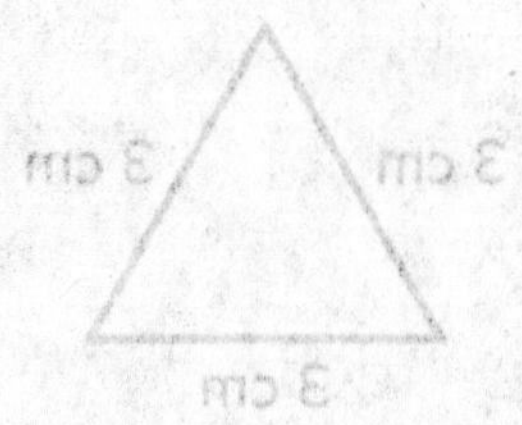

3. A triangle has an angle measure of 90°.

4. A triangle has an angle measure of 85°.

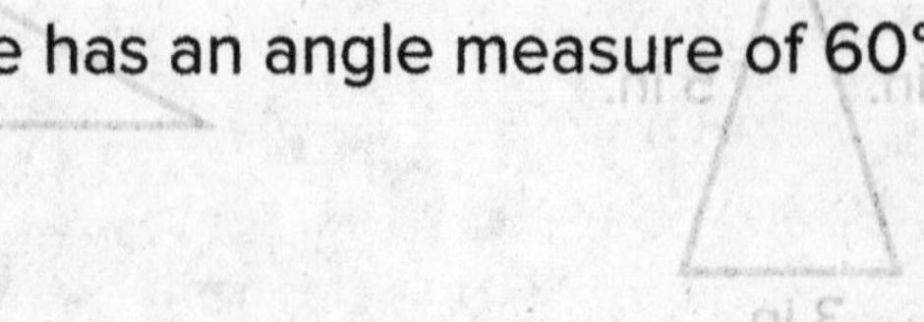

5. A triangle has an angle measure of 60°.

Understand Line Symmetry

Name ______________________________

Review

You can draw a line of symmetry to see if a shape has line symmetry.

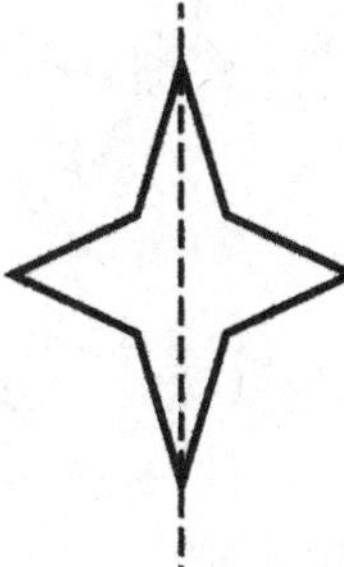

If you fold the star over the dotted line, you will have two identical halves. Thus, the star has line symmetry.

Does the dotted line show line symmetry? Write yes or no.

1.

2.

3.

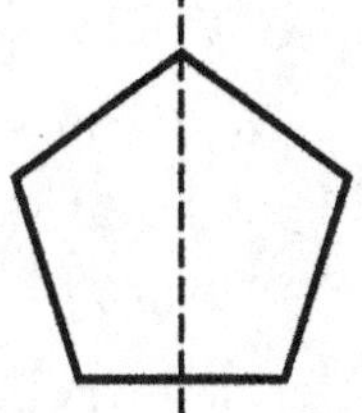

4.

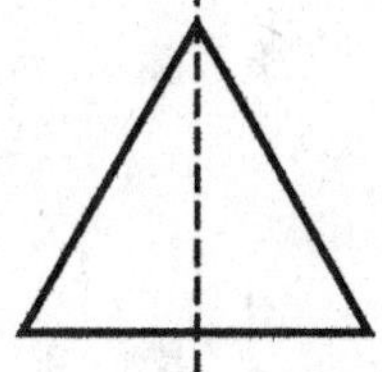

5. Write a letter of the alphabet that has only 1 line of symmetry.

Understand Line Symmetry

Name ______________________________

What is a real-world object with the stated number of lines of symmetry? Draw the object. Example: A stop sign has 8 lines of symmetry.

1. 0 lines of symmetry

2. 1 line of symmetry

3. 2 lines of symmetry

4. 3 lines of symmetry

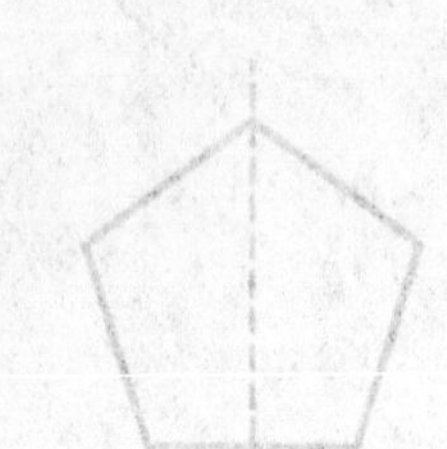

5. 4 lines of symmetry

Draw Lines of Symmetry

Name ______________________________

Review

You can use attributes of a shape to draw lines of symmetry.

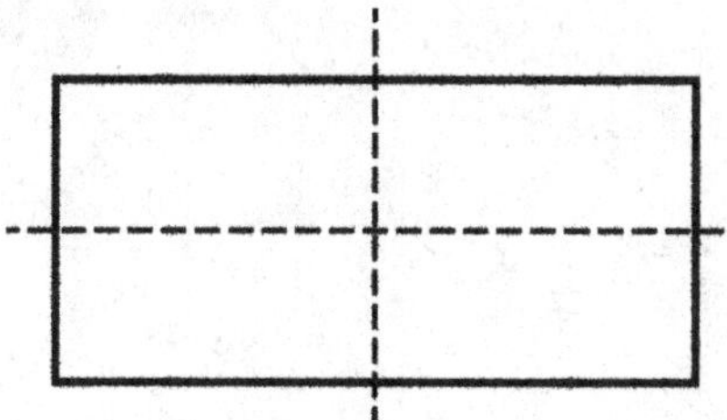

A rectangle has opposite sides of equal length and opposite parallel sides. So, a rectangle has 2 lines of symmetry.

You can draw lines of symmetry to divide the opposite sides into segments of equal length.

How many lines of symmetry does each shape have? Draw the lines of symmetry.

1. 

2. 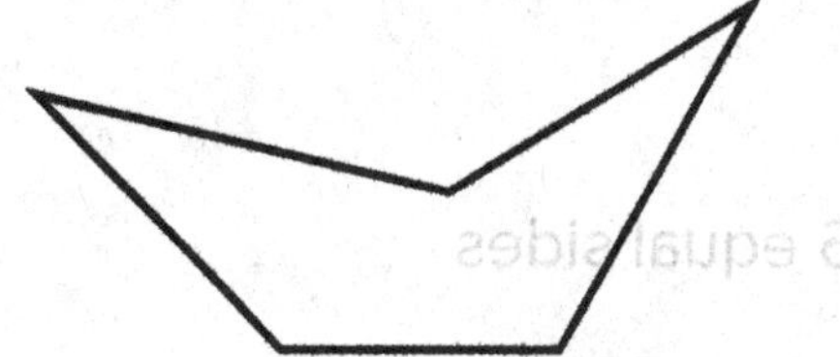

3. 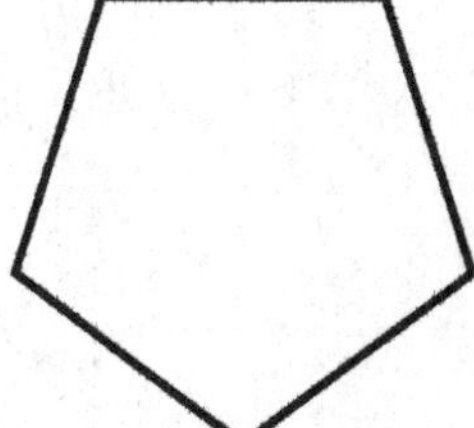

4. 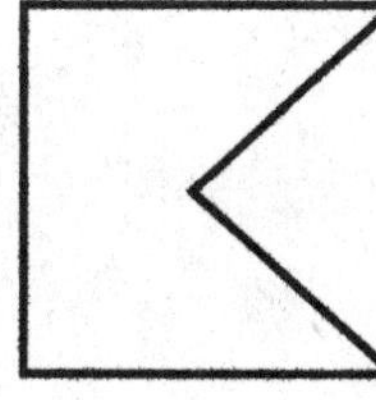

5. Draw lines of symmetry on the square stamp.

Draw Lines of Symmetry

Name ____

What shape meets the description? Draw the lines of symmetry.

1. 3 equal sides

2. 4 equal sides

3. 6 equal sides

4. 8 equal sides